"if I had known
I would have been
a locksmith"
einstein
HARROW
SOCIETY

Prepare for war – to keep the peace

Some talk of Alexander, and some of Hercules,
Of Hector and Lysander and such great names as these,
But of all the world's great heroes, there's none that can compare
With tum-tiddy-tum-tiddy-tum-tiddy-tum, to the British Grenadiers!

What's your idea of a hero? Someone who wins lots of battles? Nearly all the great names of history were successful warriors whereas little credit is given to those who avoided war altogether. Of course nowadays we don't idolize our military leaders quite so much (at least, not in Britain!). But we still give them all the weapons they say they need. Indeed, we devote more resources, time and effort to weapons today than at any other time in history.

World stockpiles of nuclear weapons are equivalent to about 50 tonnes of TNT for every human being alive today. This figure is still increasing and new weapons continue to be made, even though there are already enough to kill us all several times over. Britain spends four times as much on military research as on medical research and far more on 'defence' than on education. One in five scientists is directly employed on military work. What do they do?

Newspapers rarely mention this ceaseless preparation for war. Maybe this is because most people seem to close their minds to the thought of war—or perhaps it is the reason why they do. Yet, almost a million people in Britain are directly involved in the arms race, earning a living by the development or manufacture of weapons. Many more are indirectly involved, for the new weapons include rockets, aircraft and computers. People make the electrical and mechanical components for these weapons without knowing or thinking about their ultimate destination. And all of us, whatever our occupation, help to pay Britain's £7000 million defence bill.

We are glad that Britain has not been involved in a major war since 1945 and hope that peace will be preserved. But can we claim that the world is really at peace and that this arms race has prevented war? Since 1945 there have been over a hundred 'small wars' with Britain fighting in several. More than five million people have died—about half the number killed in the First World War (1914—18). The bombing of Korea—in which Britain participated—was greater than that of the Second World War (1939—45), whilst Vietnam received an even heavier pounding from 1966 to 1972.

Meanwhile, despite some measures of arms control, the main arms race between the United States and the Soviet Union (USSR) continues. From 1960 to 1970 the US built 2000 missile-launched nuclear warheads capable of hitting the USSR. In 1980 America had ten times as many—and the Soviet Union was not far behind. The pace of the arms race makes weapons quickly obsolete—so they are then sold to 'third-world' countries who feel obliged to keep up with their neighbours. Over three-quarters of the arms for the 'small wars' were supplied by Britain, France, the US and the USSR. So military expenditure is increasing in all countries—at a time when the poorer nations desperately need to spend more on health and education.

When the first atomic bomb exploded in 1945, it was thought by many to be the ultimate weapon. In reality it proved merely the first of a whole new generation of weapons of mass destruction. The arms race since 1945 has been even more furious than the one that led to the Second World War. An increasing number of 'small wars' have taken place and have been used as the testing grounds for the new weapons. This book describes some of the preparations for war, and discusses whether these will keep the peace.

Soft-sell hardware.

Hiroshima : the nuclear age dawns

Daily Express reporter Peter Burchett arrived in Hiroshima thirty days after the explosion and reported:

People are still dying, mysteriously and horribly—people who were uninjured in the cataclysm—from an unknown something that I can only describe as the atomic plague.

I could see about three miles of reddish rubble. That is all the atomic bomb left. . . . In these hospitals I found people who, when the bomb fell, suffered absolutely no injuries, but are now dying from the uncanny after-effects. For no apparent reason their health began to fail. They lost appetite. Their hair fell out. Bluish spots appeared on their bodies. And then the bleeding began from the ears, nose and mouth.

'At first, the doctors told me, they thought these were the symptoms of general debility. They gave their patients Vitamin-A injections. The results were horrible. The flesh started rotting away from the hole caused by the injection of the needle. And in every instance the victim dies.

Most people in Hiroshima died at once from the blast or fire. Others died later from radiation sickness. The people Peter Burchett saw dying were the first victims of this new and previously unknown illness. Yet even today people are still dying from the after-effects of the Hiroshima bomb. In the twelve months to July 1973, a further 1381 Hiroshima victims died from the effects of radiation.

Moreover, radiation damage can be transmitted to the unborn children of the victims—Hiroshima still produces higher-than-average numbers of deformed and still-born children. So it may take several generations before the effects of the Hiroshima bomb are no longer felt. The final death toll, including the after-effects of radiation, may well exceed 200 000.

The ultimate weapon?

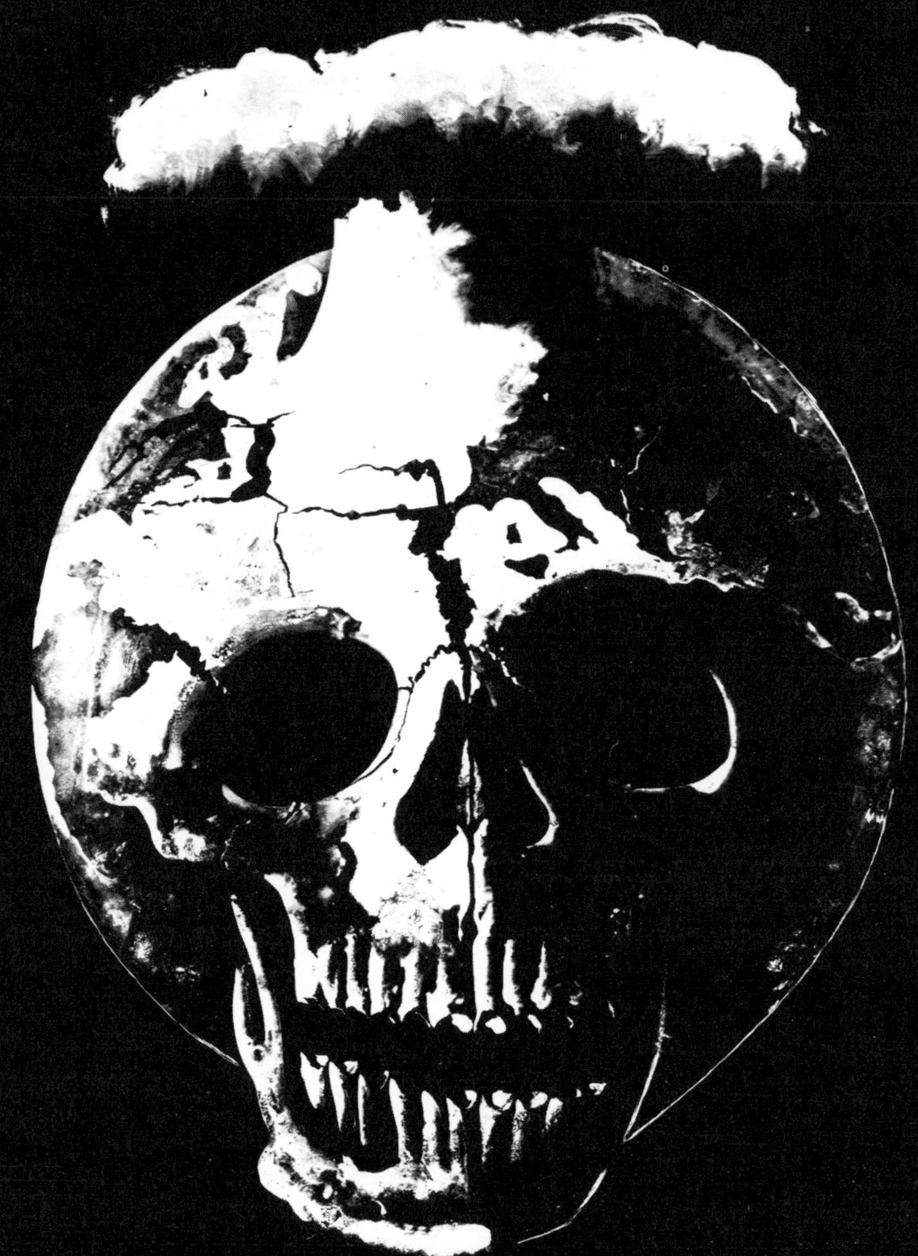

'A world war in this day and age would be general suicide.' Sir John Slessor

In 1945 the atomic bomb was hailed as ‚'The Ultimate Weapon'. Atomic bombs could totally devastate a country. People predicted,

- new weapons would not be needed
- war was impracticable against atomic bombs
- no-one would risk starting another war.

So, it was argued, the arms race would stop and war would be abolished. This turned out to be wishful thinking. All three predictions proved wrong.

New weapons have been developed. A modern thermonuclear bomb can be a thousand times more powerful than the Hiroshima bomb. A mere pin-head of a man-made poison could kill everyone alive today. Military strategists talk of 'Doomsday' and 'Overkill'.

*These weapons are *too powerful* for ordinary warfare. So they have not yet been used in an all-out war. They cannot quell a riot, protect a convoy of advancing troops, defend a bridgehead or cover a retreat. Nor is much gained by laying a country to waste: this does not help to capture its wealth or 'free' its people to live happier lives.

*There have been *more wars* since 1945 than ever before. This is partly because weak countries feel more able to tackle stronger powers, as the strong countries feel less sure about risking all-out war. For example, Soviet and Chinese arms found their way to insurgents in Algeria, Vietnam, Cyprus and Portuguese colonies without precipitating a nuclear confrontation with, respectively, France, America, Britain or NATO.

Although there has been no *nuclear* war since 1945, the existence of nuclear weapons has had a big influence on world events. We nearly had a nuclear war in 1962 over Cuba and, earlier, during the 1949–53 Korean War when US President Truman had to restrain his military advisers. General Nathan Twining fumed, 'If it were not for the politicians I would settle the (Korean) war in one afternoon by bombing Soviet Russia.'

Of course Truman was quite right. Even if the world could have survived the inevitable all-out nuclear war, this would not have done much good to Korea. The aim of war is to capture or protect things or

1962: Soviet Premier Khruschev and US President Kennedy resolve their differences over Cuba.

people. That is why smaller and more selective weapons (and even persuasion and diplomacy) are usually more appropriate than nuclear weapons.

Non-nuclear weapons also seem more 'acceptable' —the Americans were even praised for 'restraint' whilst dropping the explosive equivalent of a Hiroshima bomb a week on Vietnam! Nuclear war is so terrible to contemplate that it makes all other warfare 'conventional' and, by implication, 'acceptable'. So the atomic bomb has not stopped the arms race or ended war. Quite the reverse: it has provided a form of moral justification for new and more terrible weapons and warfare.

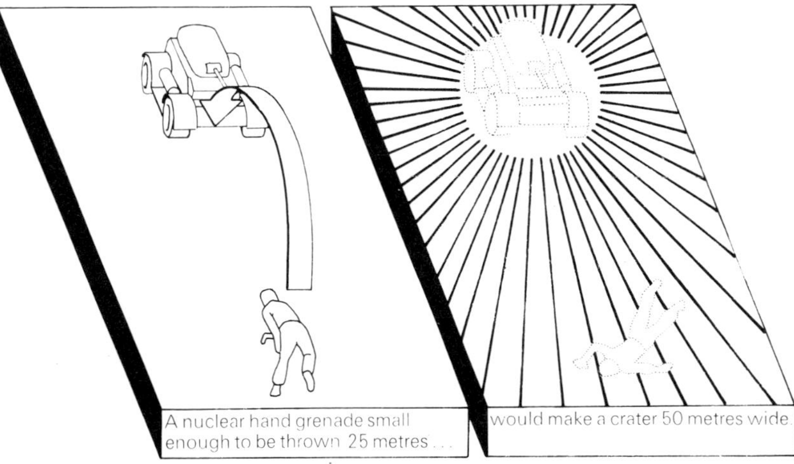

Nuclear weapons are too powerful for ordinary warfare.

War by pollution – career opportunities

This is the official symbol of Porton Down. The Latin motto means 'Watch out'.

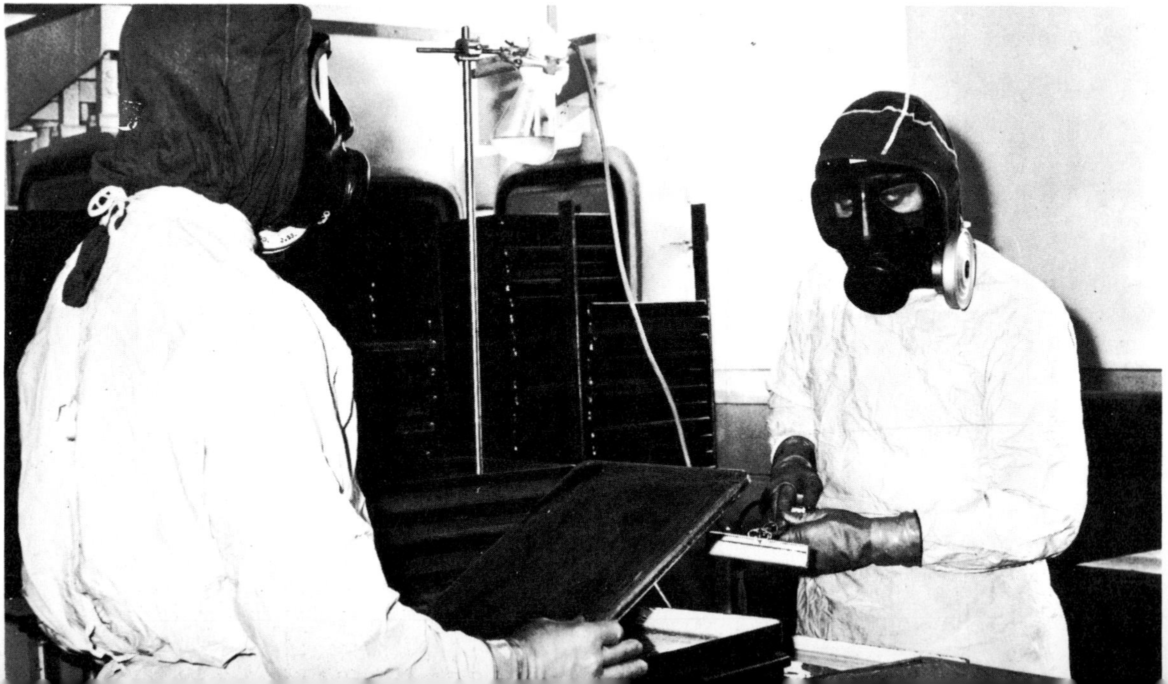

Workers at Porton Down growing anthrax. The photograph was taken at a Press Day.

Aldermaston is known throughout the world as the centre for British research into atomic weapons. Less well-known, but some would say more important, is Porton—the centre for British research into chemical and biological weapons. Nearly all of British research on biological warfare takes place there at the Microbiological Research Establishment (MRE)—which also houses the world-famous Chemical Defence Experimental Establishment (CDEE). The two stations employ hundreds, cost millions and control 7000 Wiltshire acres for their trials. An out-station is maintained at Nancekuke, Cornwall, producing chemical agents which are delivered to Newdigate, Surrey to be packaged for military use.

Despite all this, Britain officially doesn't approve of CBW—the Porton work is merely for 'defence'—and we are simply preparing ourselves for what others may do. However, to recover some of the costs of MRE and CDEE, the British government does sell know-how (and sometimes chemicals) at a profit to less squeamish countries. And, as part of our liaison with the Americans, any work at Porton may be used by our American allies.

The research facilities at Porton are excellent and there have been no major accidents such as the Americans have experienced at their Dugway establishment where in 1968 a cloud of nerve gas escaped and drifted 25 miles. The pollution of Gruinard Island off the Scottish Coast, which was sprayed with anthrax in 1942 and will be uninhabitable for a century, is the result of a 'successful' experiment in germ warfare. Such breakthroughs show that Britain leads the world with respect to CBW. Many promising scientists have found at Porton exactly the right atmosphere for disinterested high-level scientific research.

A MESSAGE FROM THE
SECRETARY OF STATE FOR DEFENCE

A great deal has been said and written about the nature of the work at the Porton Establishments. The Open Day at the Microbiological Research Establishment last year did much to kill the myth that secret work was being carried out on an offensive capability for biological warfare. I am glad that it has been possible to arrange an Open Day for the other important Establishment at Porton — the Chemical Defence Experimental Establishment. I believe it will help the public to understand the necessity of the valuable defensive work which C.D.E.E. is doing to meet the threat of chemical warfare and that sensible people in Britain will accept that this work is a vital contribution to our national security.

It will be a truly 'open' day and visitors will see most of the work being done at C.D.E.E., but it will be readily understood that — as in other countries — some area of secrecy must be maintained if we are not to assist the potential enemy.

Denis Healey

From a booklet given to visitors at a Porton Down open day.

Chemical and Biological Warfare (CBW)

Poisonous chemicals and germs can be sprayed in a gas, atomized liquid or as a finely powdered solid. They can be exploded from bombs and shells or delivered as an aerosol mist. Giant fans have been used against guerillas to blow a fine powder into tunnel and cave sanctuaries.

Whereas Biological Weapons (**BW**) are meant to kill, some Chemical Weapons (**CW**) like tear gases are supposed to be harmless. These chemical **harassing agents** irritate people, forcing crowds to disperse and rioters to run away. CS gas—a Porton invention—is the most well-known as it has been used in Northern Ireland and Vietnam. Unfortunately, like all harassing agents, it is lethal in the high doses used to 'flush-out' insurgents from their refuges.

Incapacitating agents are also supposed to do no harm. LSD has been tested for this purpose, to 'neutralize' enemy forces so that they can be defeated without bloodshed. However, the drugs only work as intended when the right amount is taken—a tricky operation to arrange in the middle of a battle. Large doses kill while smaller doses simply make people angry. Moreover, an incapacitating dose for soldiers can kill children, old people and unprotected civilians. So these chemicals are not as humane as is sometimes claimed.

Phosgene and mustard gases were **lethal agents** of the 1914—18 war. They are now out of favour because their smell gives warning for people to escape or don gas masks. Nerve gases are preferred because they are odourless and more toxic. V-agents—another British discovery—can kill merely by touching the skin. **Toxins** (non-contagious poisons made from living organisms) are even stronger. The botulinus toxin is a thousand times more toxic than nerve gases. It could wipe out life in a given area inside six hours—and then evaporate within the following six hours leaving the area once more habitable.

Germs (BW) are slower to act but, once caught, can be self-propagating—spreading like measles or the common cold or being carried by insects, birds, animals. Thus BW can have more far-reaching effects than CW (a military *disadvantage* if both sides suffer). The big powers are not too keen on BW for this reason—and also because micro-organisms can be easily made and stockpiled in hospital-type laboratories by countries and guerilla movements that cannot afford nuclear weapons.

Perhaps this is why the big powers agreed in 1973 to ban biological weapons. But chemical weapons have proved more difficult to ban, partly because many have peaceful application. For example, polyurethane foam is made from phosgene, and the nerve gas *sarin* is used in the manufacture of various pesticides. So a complete ban on poisonous chemicals would disrupt much of the chemical industry.

CS gas in use in Northern Ireland.

US army instructions for clearing tunnel sanctuaries are not meant to be a joke.

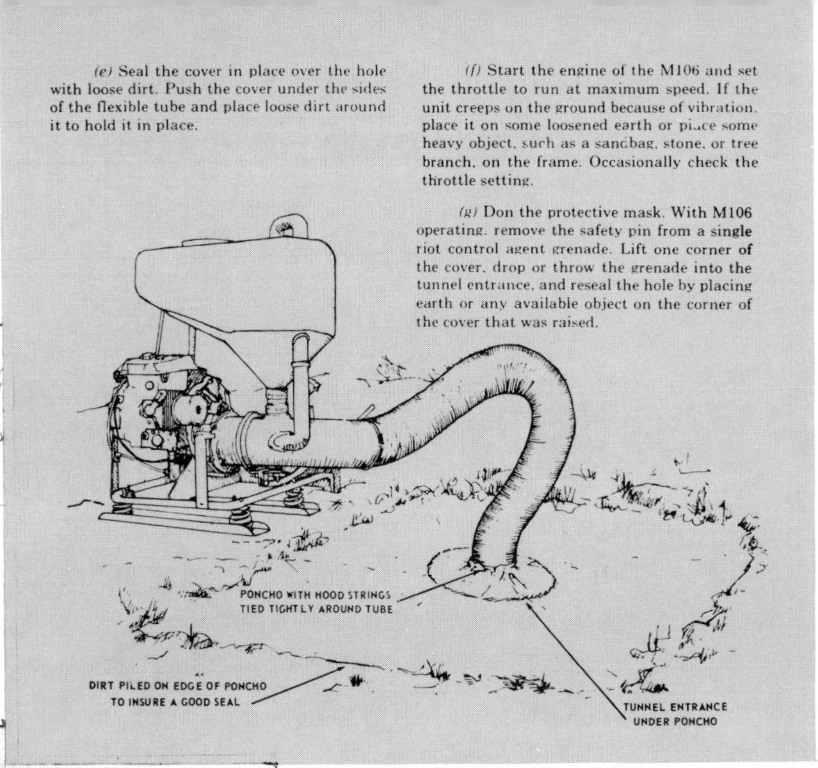

Heath Robinson caricature suggesting the use of magnets and laughing gas to clear soldiers from First World War trenches.

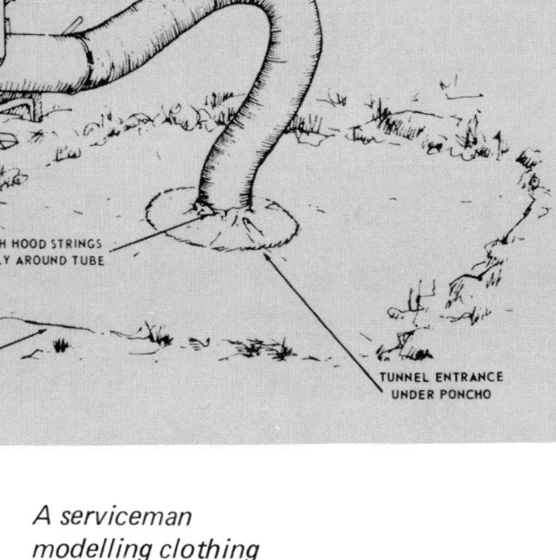

(e) Seal the cover in place over the hole with loose dirt. Push the cover under the sides of the flexible tube and place loose dirt around it to hold it in place.

(f) Start the engine of the M106 and set the throttle to run at maximum speed. If the unit creeps on the ground because of vibration, place it on some loosened earth or place some heavy object, such as a sandbag, stone, or tree branch, on the frame. Occasionally check the throttle setting.

(g) Don the protective mask. With M106 operating, remove the safety pin from a single riot control agent grenade. Lift one corner of the cover, drop or throw the grenade into the tunnel entrance, and reseal the hole by placing earth or any available object on the corner of the cover that was raised.

A serviceman modelling clothing for protection against gases at a Porton Open Day.

War makes a man of you

50. "Charge!" shouted Chang. The political instructor led the fighters out of the trench in a dash. Like tigers, they rushed at the enemy.

from 'Fighting North and South'
Foreign Languages Press Peking 1972

The fourteenth-century chronicler Jean Froissart once wrote, 'Gentle knights were born to fight, and war ennobles all who engage in it without fear or cowardice.' If he lived today he could be a scriptwriter for the army recruiting offices.

Maybe war was glorious in the past. Hand-to-hand (or rather, sword-to-sword) fighting was reasonably fair and the 'better' man stood a good chance of winning. But what sort of courage is needed to bomb from a height of 10 000 metres or more? What is the ennobling influence of Anti-Personnel Weapons?

Modern warfare is a far cry from the chivalry said to exist in the fourteenth century.

On a TV Western, a violent death is usually over in one shot—or, at the most, no longer than the victim needs to gasp out an important last message. In reality, few die instantaneously. In Alistair Maclean's novel, *When Eight Bells Toll*, he indicates how painful it would be to be 'winged' by Wyatt Earp. So there have been many attempts to draw up 'rules' for fighting 'humanitarian wars' with 'humane weapons'.

In 1139, Pope Innocent II prohibited 'the use, in Christian warfare, of the cross-bow' on the grounds that 'by reason of the very perfection of their mechanism they had become too efficiently murderous'.

In **1868** the **St. Petersburg Declaration** was signed by 17 countries and stated that the employment of arms that 'uselessly aggravate' suffering was contrary to the laws of humanity. It forbids the use of bullets which 'expand or flatten in the body or are pierced with incisions'. In other words, weapons that killed were all right but those that caused suffering, without adding to the weapon's lethal effectiveness, were not.

The **1925 Geneva Protocol** banned the use 'in war' of 'asphixiating poisonous or other gases and all analogous liquids, materials and devices'. Before the 1914–18 war, the **1907 Hague Regulations** prohibited 'poison, poisoned weapons, and weapons which kill or wound treacherously'.

Yet these agreements did not prevent the use of phosgene in the First World War nor did the Geneva Protocol prevent the use of chemical weapons in Vietnam. Once fighting gets under way the rules are ignored or loopholes discovered. (The USA was not officially 'at war' with Vietnam so the US government claimed that the Geneva Protocol did not apply!) The other difficulty is that it only needs one country to break an agreement for it to become useless: the US government refused to sign the Geneva Protocol so everyone else 'reserved the right to use CBW' against countries that themselves used CBW.

Worse still, such is the march of progress, there are now weapons expressly designed to cause human suffering. These Anti-Personnel Weapons (APW) have two main military advantages:

* Whereas dead people don't need food and medicine, the badly-wounded do.

* Whereas the dead disappear from sight, the maimed and scarred are a constant reminder of the terror of war and so sap the morale of the survivors.

Dusk Monday—3 a.m.

The Peacemaker Colt has now been in production, without change in design, for a century. Buy one to-day and it would be indistinguishable from the one Wyatt Earp wore when he was the Marshal of Dodge City. It is the oldest hand-gun in the world, without question the most famous and, if efficiency in its designated task of maiming and killing be taken as criterion of its worth, then it is also probably the best hand-gun ever made. It is no light thing, it is true, to be wounded by some of the Peacemaker's more highly esteemed competitors, such as the Luger or Mauser: but the high-velocity, narrow-calibre, steel-cased shell from either of those just goes straight through you, leaving a small neat hole in its wake and spending the bulk of its energy on the distant landscape whereas the large and unjacketed soft-nosed lead bullet from the Colt mushrooms on impact, tearing and smashing bone and muscle and tissue as it goes and expending all its energy on you.

In short when a Peacemaker's bullet hits you in, say, the leg, you don't curse, step into shelter, roll and light a cigarette one-handed then smartly shoot your assailant between the eyes. When a Peacemaker bullet hits your leg you fall to the ground unconscious, and if it hits the thigh-bone and you are lucky enough to survive the torn arteries and shock, then you will never walk again without crutches because a totally disintegrated femur leaves the surgeon with no option but to cut your leg off. And so I stood absolutely motionless, not breathing, for the Peacemaker Colt that had prompted this unpleasant train of thought was pointed directly at my right thigh.

from 'When Eight Bells Toll' by Alistair Maclean

Fire bombs

Incendiary weapons (fire bombs) incinerate their targets. They are no better than explosives against buildings but have traumatic effects on people. Few in the West have had this experience but a British major, who fought in Korea, wrote of his feelings on the matter when the photograph of Phan Thi Kim-Phuc appeared in *The Times*:

> Sir, In 1951 and 1952 I became accustomed to the daily use of napalm on Chinese positions. It wasn't until an 'accidental drop' occurred on my own company position that I realized what a nauseating and utterly terrifying weapon it is. To my mind it is a completely uncivilized method of warfare and it would be no more uncivilized to tie our prisoners of war to a post, throw petrol over them and set them alight.

Yet there *are* advantages in burning people alive. A major in the US Air Force, who also served in Korea, commented that, 'The enemy didn't seem to mind being blown up or shot. However, as soon as we would start dropping thermite or napalm in their vicinity they would immediately scatter.' His observation is confirmed in the US Field Manual, *Ground Flame Warfare*, which makes clear that people are the main targets of fire bomb attacks: 'The basic objective of fire bomb missions is to kill, neutralize, and demoralize. A secondary but vitally important objective is to destroy or damage vehicles, equipment . . . .'

In a napalm attack, the fortunate are completely incinerated by burning drops of napalm. Of the remaining victims a third may die within half an hour whilst, depending upon the available hospital facilities, others suffer a slow and painful death over a period of weeks. About one in five may recover to live out the rest of their lives permanently scarred and deformed.

What are these incendiary weapons? Most, like napalm, are petroleum-based. They comprise a sticky jelly oily mixture which burns even when sprayed with water--and adheres to organic substances like flesh. Others are metal-based (such as magnesium), pyrotechnic (containing oxidizing agents) and pyrophoric (such as white phosphorus). Magnesium and white phosphorus have the additional property of smouldering under the skin as the drops burn their way down to the bone.

Fire bombs are not solely anti-personnel weapons as they can be used against military targets. The next weapons described are solely and specifically intended to maim people and cause suffering.

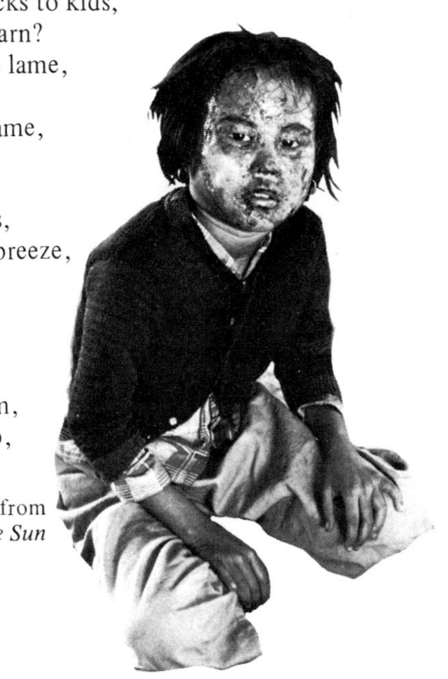

Napalm sticks to kids, napalm sticks to kids,
When'll those damn gooks ever learn?
We shoot the sick, the young, the lame,
We do our best to kill and maim,
Because the 'kills' all count the same,
Napalm sticks to kids.

There's a gook down on his knees,
Launch some flechettes into the breeze,
Find his arms nailed to the trees,
Napalm sticks to kids.

Blues out on a road recon.,
See some children with their mom,
What the hell let's drop the bomb,
Napalm sticks to kids.

US Air Force song from
The Baltimore Sun

'Blues' = helicopter gunships,
'gooks' = Vietnamese

Children running in agony from an accidental napalm drop. Phan Thi Kim-Phuc had just torn her burning clothes from her body.

...seven nations participated. The report shows an appalling increase in the use of napalm:

World War II	14,000 tons
Korea	32,357 tons
Indochina (1961-1971)	338,237 tons

The report declares that napalm is a weapon causing unnecessary cruelty and is therefore against international...

Aboard an aircraft carrier a crew member slumbers on a deadly cargo.

Anti-Personnel Weapons (APW)

Over half the bombs dropped on Indo-China were Anti-Personnel Weapons (APW). Most were fragmentation bombs, notably the Guava, Pineapple and Cluster bombs, each containing hundreds of small pellets. These devices kill or, more usually, maim people whilst leaving equipment and buildings undamaged.

When fragmentation bombs explode, pellets scatter in all directions, wounding any creatures that get in the way. The effect is similar to buckshot but with two major differences. In the first place, the pellets move at such a high speed that they penetrate flesh and enter body organs. Secondly (contrary to the St. Petersburg Declaration and all subsequent conventions), the pellets are irregularly-shaped and so expand wounds and tear at flesh.

Flechettes are arrow-shaped and so, after causing an explosive-type wound, they work through the flesh like the hook of a fishing line. Flechettes have been known to embed themselves several centimetres deep making it difficult for surgeons to remove them. A further improvement is to make them of hard plastic so that they cannot be detected by X-rays. The surgeons then have to probe blindly into

The Bat Wing mine is detonated when it is stepped on and can blow off a man's foot.

the victim's flesh to find where the flechette has gone. Yet another idea is to tip the pellets with depleted uranium causing chemico- and radio-toxic effects.

All these weapons clearly contravene the 1868 St. Petersburg Declaration prohibiting weapons that 'uselessly aggravate' suffering. They are used because they tie up valuable medical resources and occupy the time of many skilled doctors and nurses. The pellets have no effect on military targets but are very good at maiming people whilst leaving them alive.

To make sure that the pellets actually hit people, rather than just houses or trees, many are designed to be triggered by movements on the ground. The **Wide Area Anti-Personnel Mines** (WAAPM—say that fast and you'll get the idea) are the most sophisticated. They are dropped in large canisters which, after they hit the ground, throw out eight fine threads to act as trip wires (which is why the Vietnamese call them **SPIDER mines**). Any movement of the trip wires causes pellets to explode over a distance of about 60 metres. If undisturbed for a long time, the mines are detonated automatically by a time-fuse (so that it is safe for troops later on).

Some mines have been made so small that they can be used as booby traps (and, by others, as letter bombs). The **Dragon's Tooth** or **Bat Wing mine** is only 40 mm long and weighs as little as a small envelope. Like the pellets it is no use against military targets: it cannot even puncture a truck tyre. The **Gravel mines** ('**Leaf mines**' to the Vietnamese) are small scraps of brightly-coloured cloth or plastic containing explosive and, sometimes, a few pellets. They work in the same way as the Bat Wing mine and are especially effective against inquisitive children. The point of these is to make life difficult for a peasant community without endangering troops and armoured vehicles.

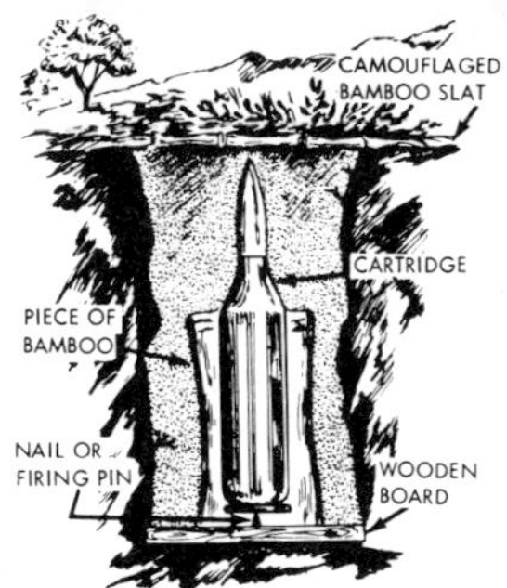

The concealed bullet trap above is known as the 'Footbreaker'. Guerillas often use such devices against superior military forces which they dare not engage in direct battle. The spiked board in the boobytrap below, swings up when you step on it.

WAP-092002-9/20/72-WASHINGTON:This photo obtained from the Pentagon shows an XM41E1 Gravel Mine dismantled. It is believed to be the same type of device used to kill an Israeli diplomat in London and also discovered by London postal workers in mail addressed to the Israeli Embassy there. The mine when assembled is about the size and shape of a teabag. The device,mailed in an envelope and fatally injuring Ami Shacori,Israel's agricultural counselor,was believed by Scotland Yard sources to have been American made. (UPI) SEE '029A

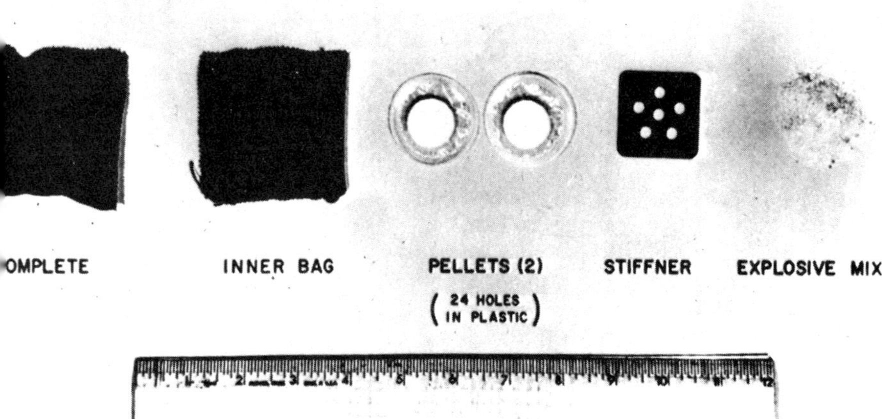

Counter-insurgency

Riot control forces in action in Belfast.

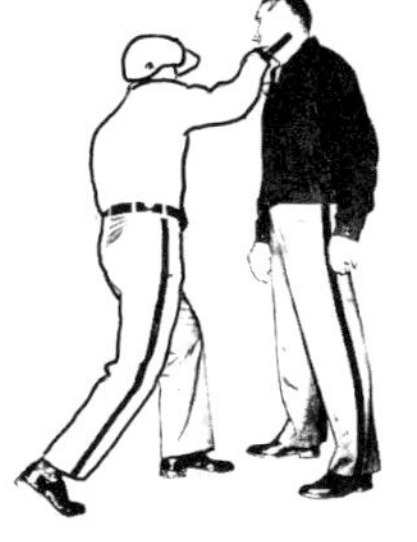

Modern warfare hurts civilians just as much as the professional armed forces. This was not always so. In the past ordinary people were less concerned and might even be unaware that their king was off on another war. Very often it didn't matter much who won: life for a peasant or farmer went on more or less the same whoever became their overlord.

Nowadays people are much more concerned. This is partly because modern warfare is far more destructive and involves us all, whether we like it or not. In addition, many present-day conflicts are *caused* when people try to change their governments. When political action fails to achieve a change, armed conflict follows —sometimes aided by arms supplied by political allies in other countries. This type of conflict inevitably involves large sections of the civilian population, for or against the insurgents.

Brigadier Frank Kitson is a specialist in 'counter-insurgency'. Subversion and insurgency, he says, differ from other forms of warfare mainly because:

'force, if used at all, is used to reinforce other forms of persuasion' (like strikes and protest marches) 'whereas in more orthodox forms of war, persuasion in various forms is used to back up force. . . . the qualities required for fighting a conventional war are different from those required for dealing with subversion or insurgency. . . . Traditionally a soldier is trained and conditioned to be strong, courageous, direct and aggressive but. . . . they often find that these good points are exploited by the enemy. For example, firm reaction in the face of provocation may be twisted by the enemy in such a way that soldiers find the civilian population regarding their strength as brutality. . . .'

This type of warfare needs special weapons. Non-lethal weapons are used for riot control—to disperse crowds and demonstrators without injuries. In practice of course, this is difficult to achieve. There are many people in Northern Ireland suffering today from the effects of CS gas and rubber bullets, and the use of these non-lethal weapons often inflames passions so that conventional weapons are used in retaliation.

If everyone was prepared to accept government decisions, and governments were prepared to give up unpopular policies without using force or suppressing democratic debate, then there would be no insurgency or counter-insurgency. As it is, even if complete and general disarmament were to be agreed between governments, some armed forces would be desired by the governments to keep themselves in power!

Warning leaflet issued by Security forces in Ulster.

Terrorists and guerillas

When a politician speaks of a 'bomb outrage' you can be pretty sure he is referring to an incident where two or three innocent bystanders were wounded or killed, not the systematic destruction of hundreds of towns and villages. When he refers to an act of violence and blackmail he may mean a kidnapping attempt but never the mass deportation or resettlement of suspects. In the language of the politicians, the 'men of violence' are terrorists, not police or armed forces.

Why do you think terrorists get so much publicity? More people have died since 1945 from *tests* of nuclear weapons than from acts of terrorism. Is it because the politicians are themselves threatened by terrorism but not by war? Whatever the reasons, the professional armed forces—despite their APW and CBW—hardly compete with the amateurs when it comes to publicity.

Sometimes terrorists resort to violence because they lack political support. Individual acts like assassinations are employed by groups who have little regard for the views of others and who might well, were they to win power, themselves rule by force. But, more commonly, terrorists do have a measure of popular support and are reacting (possibly unwisely) to the violence of their oppressors. So it is important not to condemn the political and social aims of a terrorist group purely because you dislike their methods. Like every political movement they are attempting to right social wrongs and their solutions should be judged on merit.

Car bomb at the Old Bailey, London, 8 March 1973.

When a political movement has wide support, but democratic channels are closed, it rarely bothers with terrorism (which often kills the innocent and alienates others). Instead it organizes an armed insurrection and engages in guerilla warfare. In a guerilla war, one side—the established authority—has much stronger armed forces, but the other—the guerillas—has strong political support. This means that the guerillas dare not risk orthodox battles: the classic guerilla tactics are 'hit and run', immortalized by the four 'rules' of the Chinese Red Army led by Mao Tse-Tung

1. When the enemy advances, we retreat!
2. When the enemy encamps, we harass!
3. When the enemy evades, we attack!
4. When the enemy retreats, we pursue!

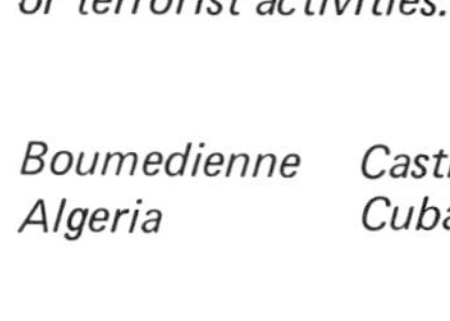

Boumedienne   Castro
Algeria   Cuba

Mao   De Valera
Tse-Tung   Eire
China

Part of a Chinese cartoon story starring woman guerilla Chao Yu-Min.

To counter these tactics, the government forces need to sever the links between the guerillas and the local population. If this cannot be done by persuasion, the local population may be 'resettled' elsewhere. If this is impracticable, more powerful weapons may be deployed. Anti-personnel weapons, defoliation and carpet bombing are examples of the techniques used to deny guerillas sanctuary.

Terrorism thrives when letter bombs gain more attention than letters of protest and when conventional political action is not possible. Leila Khaled is a Palestinian Arab, evicted from her homeland at the age of four. She is unable to return home to argue her case and believes that terrorism is, for her, a justified method of protest:

'I do not see how my oppressor could sit in judgement on my response to his oppressive actions against me. He is in no position to render an impartial judgement or accuse me of air piracy and hijacking when he has hijacked me and my people out of our land. If the enemy defines morality and legality in his own terms and decides to apply his ethical and legal doctrines against me because he has the power. . . . I am under no obligation to listen, let alone obey his dictates . . . . My deed cannot be evaluated without examining the underlying causes.'

Ecocide

Modern warfare has had far-reaching effects on landscapes and on the ecology of the battlefields: indeed, war has probably harmed our environment more than industrial pollution. Professor Barry Commoner, one of the best-known writers on environmental matters, has commented on this strange blind spot. 'There are only two ways to ruin our environment, carelessness and deliberate intent. Why is there so much concern about the former when the damage from the latter is so much greater?'

Just as 'genocide' refers to large-scale intentional actions to destroy human beings (i.e. mass murder), 'ecocide' (a word yet to appear in most dictionaries) refers to large-scale intentional measures to destroy the environment. In the long run, 'ecocide' could kill more than 'genocide': it may involve the permanent destruction of arable land and forests, depriving future generations of food and shelter.

It is not new for crops to be destroyed in a war. In the Middle Ages, and earlier, towns were beseiged until food supplies were exhausted and, to speed the surrender, local crops were burnt. But it was never the practice to destroy the land in such a way as to prevent its recovery for subsequent seasons. For this 'advance' to be possible, the resources of modern technology were needed.

Defoliation

During the Vietnam war, a very powerful combatant, with unchallenged air superiority, tried for years to dislodge an enemy living with and amongst the local population and hiding in an extensive network of underground tunnels. Steps were taken to deny the guerillas food and shelter. Over 50 000 tonnes of herbicides were dropped on forests and agricultural land in 'cover-denial' and 'food-denial' operations.

'Cover-denial' meant stripping leaves, foliage and vegetation to remove cover protecting guerillas and their supporters. 'Food-denial' killed the crops so that local farmers could no longer supply the guerillas and, just as important, would need help from outside to avoid starvation. The herbicides were applied at about ten times the 'safe' rate recommended for agriculture and as a result drinking water was contaminated so that babies were born dead or deformed.

Approximately one eighth of the land of South Vietnam was sprayed from 1960 to 1970, an area as big as Wales. Food for more than a million Vietnamese, and timber equivalent to 30 years' needs, was destroyed. This damage to the vegetation, the terrain and the environment was intentional and recovery may take decades. So it seems apt to describe the action as 'ecocide'. The herbicides programme only ended in 1970 when it became clear that other methods were more effective.

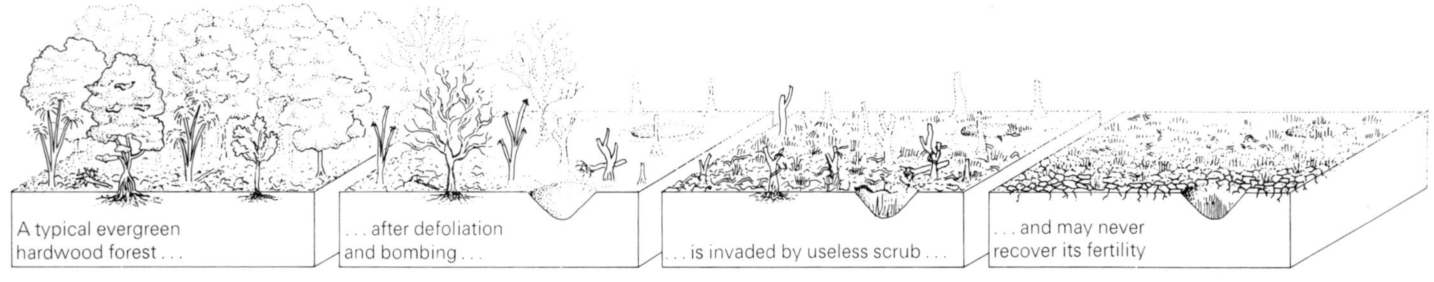

A wide strip of defoliated forest in Vietnam.

The bombing of Indo-China shifted more soil than the building of the Suez and Panama canals combined.

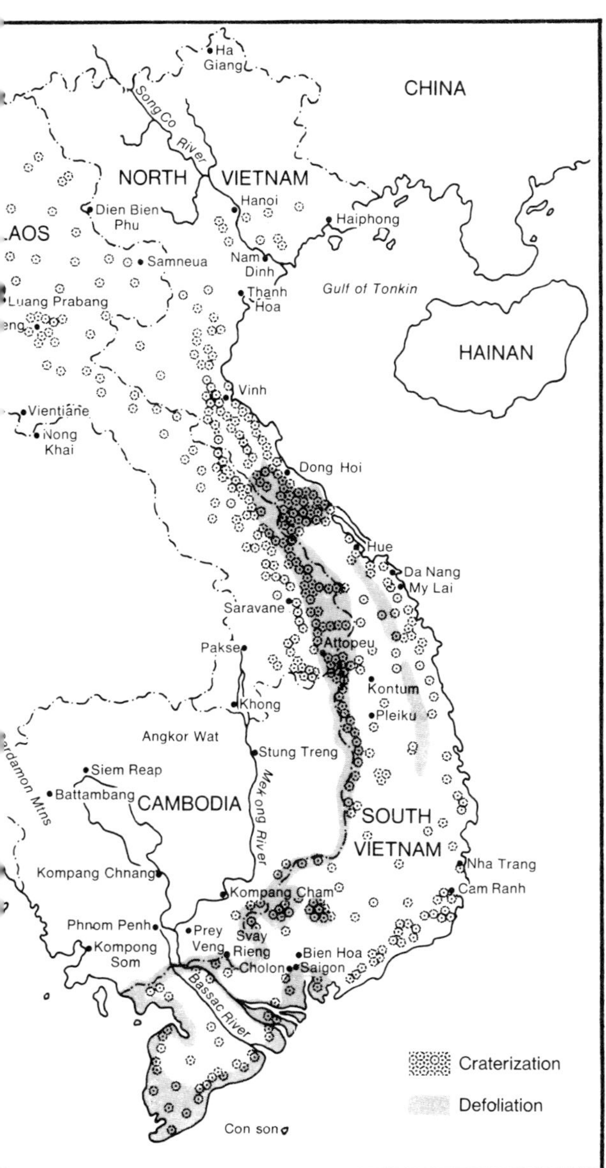

*Not exactly precision bombing
(the Second World War).*

*A munitions factory in the First World War—
the 'war to end all wars'.*

'Landscape management'

By the late 1960s it became clear that the herbicides programme was expensive, not particularly effective (in military terms), and that side-effects were endangering both friend and foe. New techniques were tried to deny the guerillas food and sanctuary.

Land clearing

To lessen the danger of ambush, military roads are often built with wide verges. In Vietnam the idea was carried further to give 'protection' to towns and villages. The devastation was enormous. Vegetation was cleared for 200 metres on either side of highways and around villages, leaving the land in the churned-up state reminiscent of motorway construction. An area bigger than the whole of Greater London was scraped bare.

It probably will be a permanent wasteland because, deprived of the protective cover of trees and vegetation, tropical soil hardens when exposed to sunlight. Sometimes it becomes rock solid and never again may be used for agriculture. Moreover, as villages are mostly located close to arable land, a high proportion of the best and most fertile land has been lost through this land clearing.

'Conventional' bombing

Ordinary bombing raids make their own contribution to ecocide—even though trees and plants are not in this instance the primary targets. All the same, an estimated 25 million bomb craters caused permanent ecological damage to the landscape of South Vietnam.

Vast tracts of formerly fertile land were now left untended. Irrigation schemes were spoiled and, near the coast, sea-water encroachment has resulted. New breeding ponds were created for mosquitoes carrying malaria, dengue and haemorrhagic fevers. Thousands of trees were torn up and shrapnel damaged millions more, inviting fungal infection and wood rot. It will be years before the effects of this incidental 'landscape management' are erased.

'Carpet' bombing

Life is not all that easy for the crew of a bomber aircraft. If they fly too low they may be hit by ground fire. If they fly too high they are likely to miss their targets. 'Carpet bombing' enables targets to be destroyed from a great height in relative safety (for the airmen).

On a carpet-bombing mission, several planes take part, flying close together, almost touching wing tips, releasing bombs at set intervals (usually every second or so). The bombs fall in a broad 'carpet' for many kilometres, cutting jungle roads, killing any creatures who happen to be underneath. This has all the disastrous environmental effects already mentioned for defoliation, land clearing and conventional bombing. In addition, forests are divided into sections making it still more difficult for the eco-system to recover.

Vegetation takes years to recover from this kind of assault.

A clean and impersonal war

Although they never saw the carnage, the crew that bombed Hiroshima did know that they were destroying a city. In a future war, pilots—if they still exist—will not need to know anything about their targets. These will be coordinates on a map, not towns or villages.

Since 1970, many targets in Indo-China were chosen by computer. The pilots simply flew over a selected route and the bombs were dropped by remote control from instructions radioed from the computer centre of the ground control base. The computer worked out these instructions from information relayed by 'sensors' dropped into the general region of the target prior to the attack.

Inevitably mistakes were made, especially since the resourceful Vietnamese found ways of confusing the sensors—for instance, hanging bags of urine in trees to overload the 'people-sniffers'. Despite these teething troubles, technological progress continues—pilotless aircraft are on the way—and the need for human beings is fast diminishing.

This has big military advantages. Bravery becomes automatic, loyalty unquestioning, action instantaneous and responsibility lost amongst the printed circuits. The robots that do the fighting will not even care who wins! In 1969 General Westmoreland, the US Army Chief of Staff gazed dreamily into the future:

On the battlefield of the future, enemy forces will be located, tracked and targeted almost instantaneously through the use of a data-linked, computer-assisted intelligence evaluation and automated fire control.

With first round kill probabilities approaching certainty, and with surveillance devices that can continually track the enemy, the need for large forces to fix the enemy physically will be less important.

I see battlefields or combat areas that are under 24 hours real or near-real-time surveillance of all types .... I see battlefields on which we can destroy anything that we can locate through instant communications and the almost-instant application of highly lethal fire power.

With cooperative effort, no more than ten years should separate us from the automated battlefield.

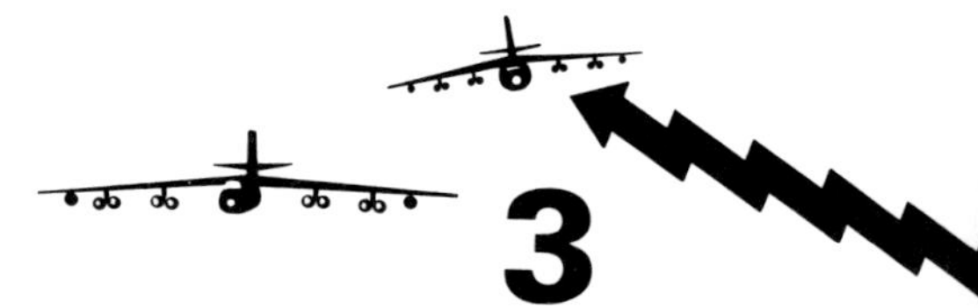

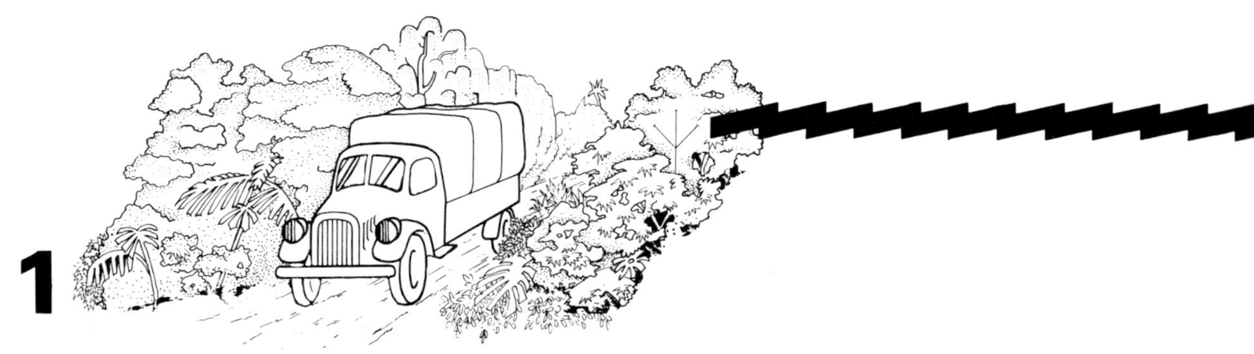

FLOYD My name is Jon Floyd. I was a pilot based with Marine Attack Squadron 533. We had access to sailboats and mostly spent our time in the Officer's Club and nominally doing some sort of job. We also flew into North Vietnam. . . . What we did, while I was there, was going in at night, acquiring a target on radar, and through the information from the various systems, it went into a computer; I pulled the commit switch, and the computer dropped the bombs. . . . You fly. You see flak at night. That's about as close to the war as we get. Sometimes you get shot down, but you don't see any of the explosions. You can look back and see 'em, but you don't see any of the blood or any of the flesh. It's a very clean and impersonal war.

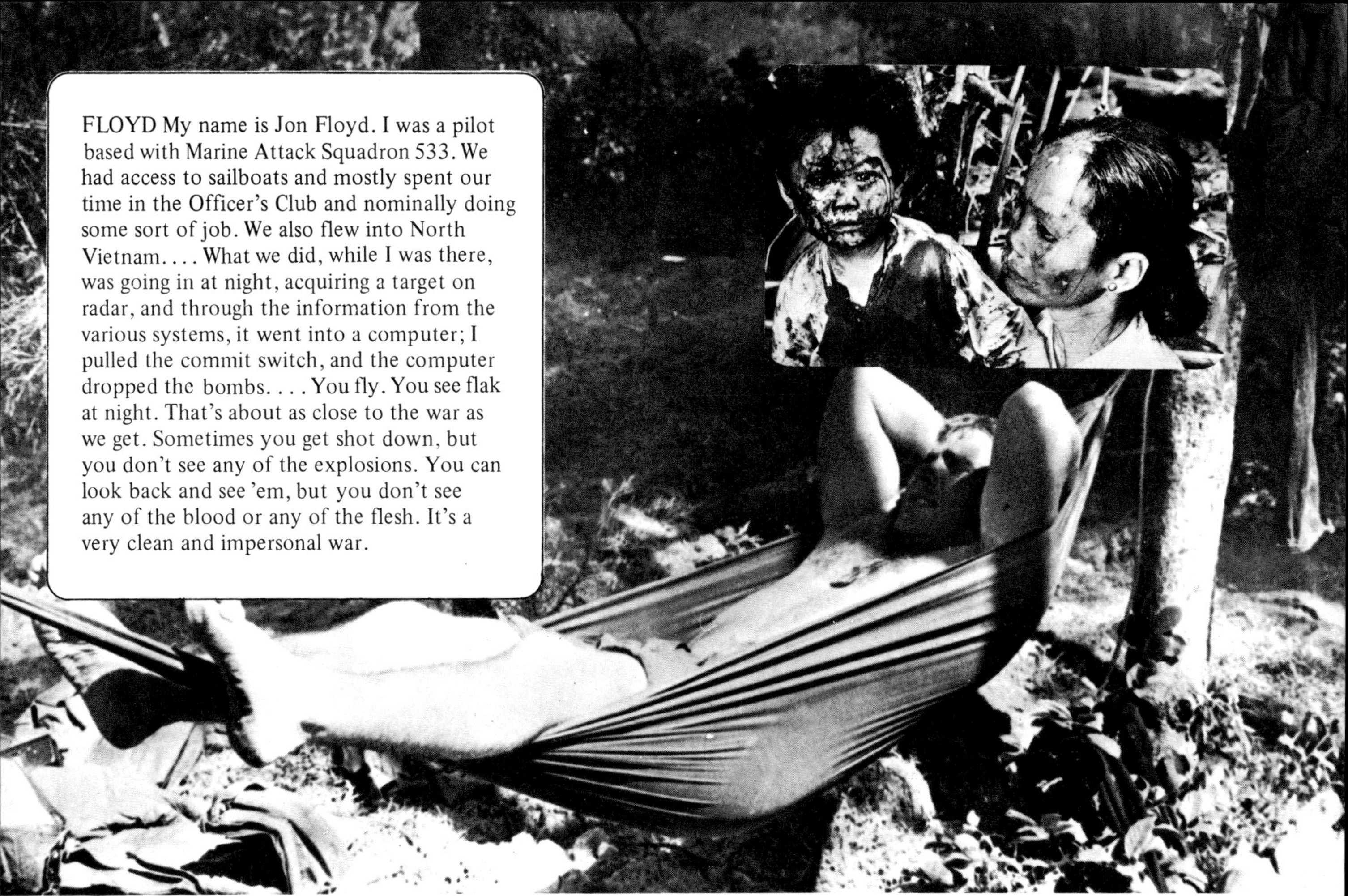

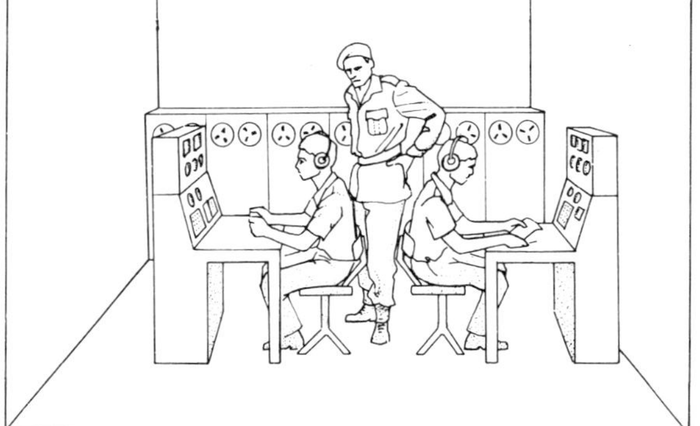

2

1 *Sensors are dropped from aircraft. Acoustic and seismic sensors detect sounds and movement. Other sensors detect temperature and humidity changes and recognize body odours. This information is radioed back to the control base.*

2 *Far away at the control base, experts listen in and analyze the information. If they decide that they have located enemy forces, authority is given to the computer centre to plan an attack.*

3 *Bombers fly over a route selected by the computer and drop their bombs according to its instructions.*

Seven H-bombs on Britain

Effect of a ground burst of a 1 megatonne bomb on a city of about 1 million

Key for each 10 megatonne explosion

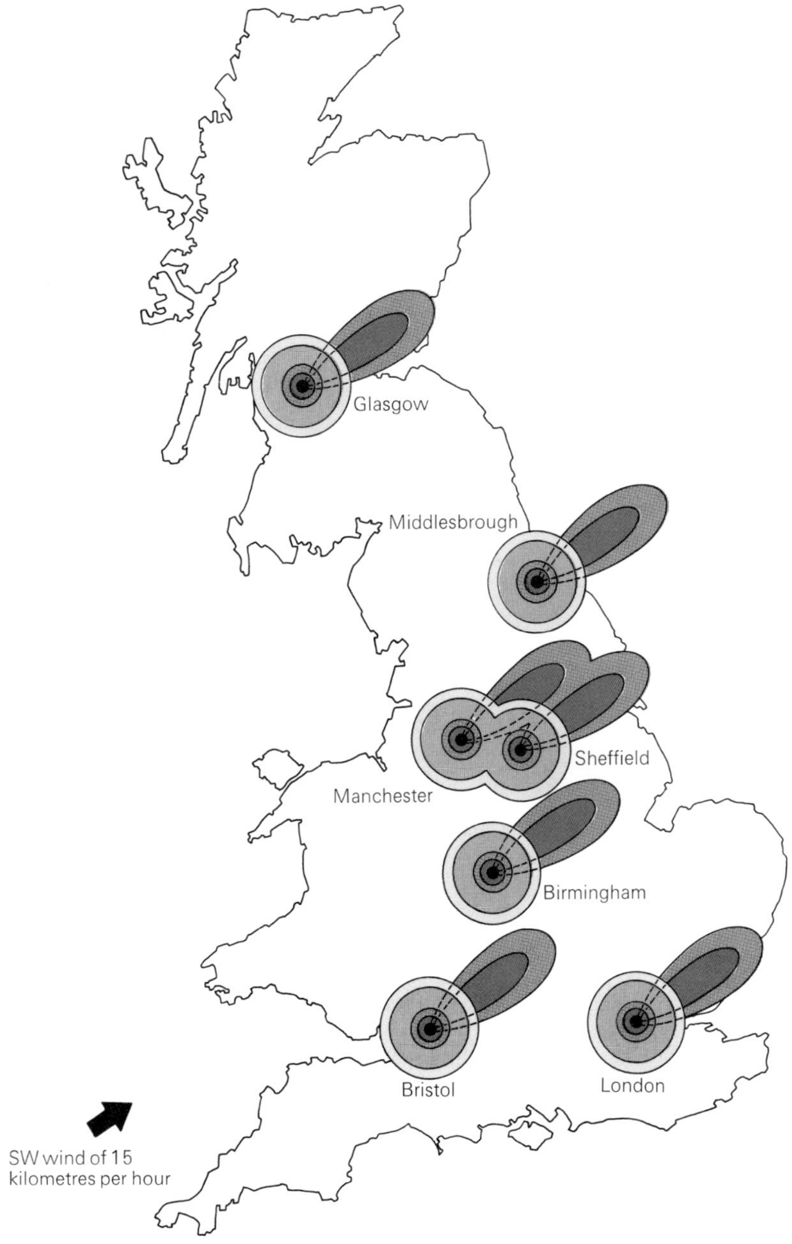

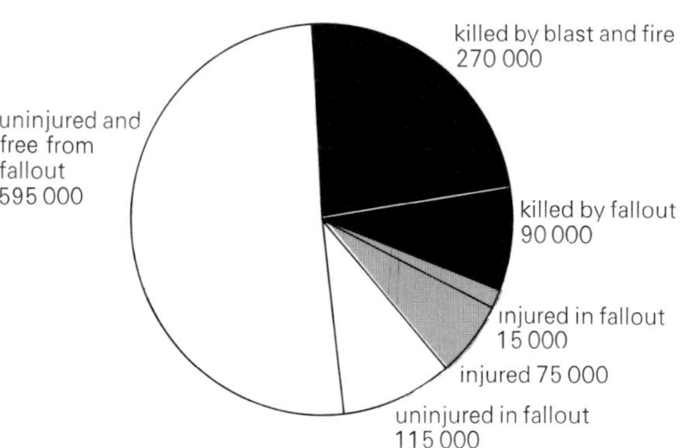

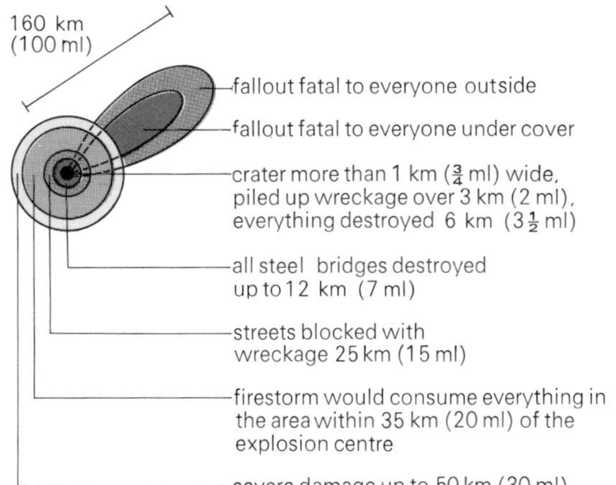

Carpet bombing and defoliation are all very well against ill-equipped guerillas. In a nuclear war there would be no time for this leisurely approach. Thermonuclear bombs do the job more quickly and effectively. The map shows what might happen if 10-megatonne bombs were dropped on seven big population centres in Britain. About 15–20 million would be killed at once from the effects of blast and fire. Many more would die from radioactive fallout in the areas downwind from the explosions. More again would suffer from the long-term effects of strontium-90 and other long-lived radioactive substances.

So what would happen to Britain in a nuclear war? Strangely, we might be better off if we lost badly! This is because most British (and American) nuclear bases are in East Anglia, which is relatively under-populated, and Scotland. If our opponent attacked first and destroyed these bases, he would not need to bother to attack the rest of Britain. But, if we or our allies were first to attack and failed to destroy completely his bases, we would then face retaliation. The enemy would see no point in attacking the empty nuclear bases so he probably would take revenge on as many people as possible, as shown on the map.

The effect of a thermonuclear bomb depends upon the height at which it is exploded. **Exploded near the ground**, it scoops out a vast crater, smashes everything for several kilometres and forms a towering mushroom cloud of radioactive dust and debris that comes down as deadly fallout. A near direct hit could 'knock-out' an underground military base. **Exploded high up**, where the air is thin and more energy goes into heat, it is mainly a fire weapon for use against people and buildings above ground. A 5-megatonne explosion 50 kilometres up would send a searing heat wave over an area of 50 kilometres diameter.

A 10-MEGATONNE BOMB WOULD

*flash quicker than the eye can blink for protection, burning the eyes of people looking at it from 400 or 500 kilometres away
*grow within 40 seconds to a blindingly bright fire-ball more than 5 kilometres across, as hot as the inside of the sun

AND, depending on the height of the explosion,

*blow out a crater, deeper than London's deepest underground railway and over 1 kilometre wide, producing a rim of piled-up wreckage up to three kilometres wide
*smash everything within 6 kilometres, destroy all steel bridges for 12 kilometres, block streets for 25 kilometres and cause further severe damage up to a distance of 50 kilometres
*fatally burn anyone in the open (40 kilometres) and ignite fires (50 kilometres). The fires would coalesce into a 'firestorm'—a gigantic uncontrollable fire that sucks in air at hurricane force to fan and feed the flames—to consume everything in the area within 35 kilometres of the explosion centre
*kill most people within a 50 kilometre radius.

Radioactive fallout

You cannot see or feel radiation. It enters into the body and changes molecules of haemoglobin and other organic matter into noxious chemicals—a sort of internal poisoning. You then suffer from leukaemia or some other radiation-induced disease without ever noticing the actual radiation.

Radiation also can affect genes so that children are born dead or deformed. The children of radiation victims may transmit defects to grandchildren and great grandchildren. Even today Hiroshima produces higher-than-average numbers of deformed children and leukaemia victims.

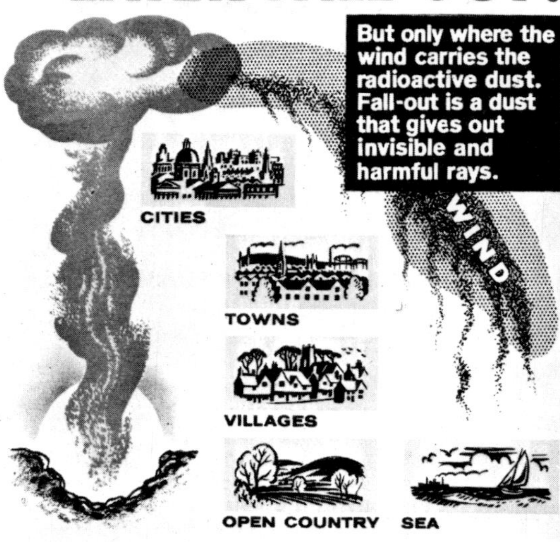

Some scientists estimate that thousands have died from nuclear radiation caused by test explosions of nuclear weapons. Yet, radioactivity from man-made explosions *on average* rarely exceeds the background radiation of nature. It is the *chemistry* of artificial radioactive substances created in nuclear explosions that constitutes the greater hazard. Many of these substances have affinities with chemicals used by our bodies. Strontium-90 is like the calcium of bones and teeth. Iodine-131 is absorbed in the thyroid gland. Caesium-137 is similar to potassium and carbon-14 virtually identical to ordinary carbon. These chemicals *concentrate* in living matter because of the way living creatures feed upon each other.

To illustrate this point, here is an example documented for the discharge from a nuclear power plant on the Columbia River. The radioactivity in the river was very small (quite safe in fact) but:

* the river plankton was 2000 times more radioactive
* the fish were 15 000 times more radioactive
* the ducks were 40 000 times more radioactive
* young swallows fed on river insects were 500 000 times more radioactive
* egg yolks of water birds were more than a million times more radioactive than the river water.

As a result of these sucessive steps, the animals high up in the food chain received the most radio-activity: a man drinking three glasses of local milk and eating a ¼lb of local beef each day would receive more radioactivity than the 'safe' limit for workers inside the power plant. And, because these substances are *inside* our bodies, all their emitted radiation affects living cells. This is why radioactive fallout is far more dangerous than natural radiation even whilst its total quantity is much less.

No-one can be sure what would happen after a nuclear war but it is known that the radiation from only 10—20% of present nuclear stockpiles could kill us all. Radioactive fallout will ensure that life will not be too good even for the 'winner' of a nuclear war!

Strontium 90

Words and music
by Fred Dallas

Every time there's an H-Bomb test,
  You get a little more strontium under your vest;
It rots your bones, makes your teeth fall out,
  And makes life shorter there's no doubt.

Chorus:
  Strontium 90, strontium 90, falling all around;
  Strontium 90, strontium 90, poisoning the ground.
  With the fall-out falling all the while,
  We'll soon be lit like a luminous dial,
  If we don't stop strontium 90 falling all around.

Drink more milk the posters say,
  Drinka Pinta Strontium every day.
But I've heard tell, and I know its true –
  That strontium ain't no good for you.

Those experts say we're safe enough;
  Takes a lot more strontium before we snuff.
But personally I've had my fill,
  'Cause the only safe dose is absolutely nil.

Leukaemia will make you sick,
  And when you get it you go pretty qick.
With the fall-out up and the blood count down;
  Just a little while and you're under ground!

If you're tired of eating atomic dust,
  Got to stop the tests or the world goes bust.
The only party that gets my vote
  Says "BAN THE BOMB" on its election note.

Strontium 90 does not exist in Nature. It is a man-made radioactive poison released by nuclear explosions.

From test-areas, however remote, it circulates round the earth and is deposited by rain. It contaminates herbage or is washed into the soil from which it is absorbed by plants.

It is eaten and drunk and, because in the living body it behaves like calcium, it is built into the bones from which it cannot be removed by any known means.

Children acquire more of it than grown-ups because they are building their bones.

Our children get most of their dietary-calcium in milk, so that much of the fallout Strontium is retained by the cow which processes the contaminated herbage into milk.

But in countries (especially the rice-eating countries) which rely on cereals for their dietary-calcium, the amount of Strontium in the bones of children is six times as high as here. This is already in excess of the figure which the British Medical Research Council said would require "immediate consideration".

Strontium 90 has a half-life of 28 years, i.e., in that period half the radioactive atoms absorbed at any given time will still be splitting. Thus the present radiations will persist into adulthood, and any further fallout will be added to the sum in the bones.

ALL INCREASED RADIATION IS HARMFUL. That is a simple axiom beyond scientific disputation or official reassurance.

Four minute's warning,
  What shall we do ?
Brown paper, whitewash,
  Out goes you.

4 PREPARING THE HOUSE

Guard against Fire

The H-bomb's heat could not set fire to the brick or stone of a house but, striking through unprotected windows, it could set fire to the contents.

Stop the heat from entering the house

Whitewash your windows; those at the top of the building matter most. The whitewash will greatly reduce the fire risk by reflecting away much of the heat, which would have passed by the time the slower moving blast wave arrived. The blast might shatter the glass, but keeping out the heat-flash for those few seconds would prevent countless fires.

These handy hints for preparing for nuclear attack are taken from two HMSO booklets—The Hydrogen Bomb, price 9p and Nuclear Attack (now out of print).

Keep buckets of water on each floor

A stirrup pump or garden syringe would be very useful. If you have one, test it and make sure that it works. Keep it close to your fall-out room.

Guard against Flying Glass

Draw curtains and keep blinds lowered and closed. This will help to prevent injury from splinters of flying glass if the windows were shattered by the blast.

The stairs would give some protection against falling debris

A slit trench with earth covering protects against blast and radiation

If there were no cellar, the room with fewest outside walls would make the best indoor refuge

Civil Defence

Could we survive a nuclear war? Some people think we might. That is why we have **Civil Defence**. The government has built underground shelters in semi-secret locations. The idea is to protect about one in a hundred from the initial blast and fire. After a few weeks or months, these survivors would re-emerge to start life again.

Who would be chosen for this protection? It is rumoured that lists included local police, council officials, the vicar and a duchess—as well as the expected government and military chaps. It needs little imagination to see that this selection would be insufficient for the continuation of the British people!

Perhaps these Civil Defence preparations are window-dressing, to give foolish people some hope of survival. At the time of the 1962 Cuba crisis, the British government decided not to alert the Civil Defence forces or issue any instructions in case this caused a panic. They also feared that the Russians might interpret such action as a preparation for an attack. In order not to precipitate a Soviet offensive, the government preferred to take no action to protect civilians!

In practice, even underground shelters cannot solve the problem of long-term fallout. And the animals left above ground will have suffered even more than the humans. So it will be difficult to re-establish the life cycles of modern farming. If there are any survivors they may have to revert to root grubbing and subsistence agriculture.

NOTICE

OFFICE OF CIVILIAN DEFENSE
WASHINGTON D. C.

**INSTRUCTION TO PATRONS ON PREMISES
IN CASE OF NUCLEAR BOMB ATTACK:**

UPON THE FIRST WARNING:

1. STAY CLEAR OF ALL WINDOWS.

2. KEEP HANDS FREE OF GLASSES, BOTTLES, CIGARETTES, ETC.

3. STAND AWAY FROM BAR, TABLES, ORCHESTRA, EQUIPMENT AND FURNITURE.

4. LOOSEN NECKTIE, UNBUTTON COAT AND ANY OTHER RESTRICTIVE CLOTHING.

5. REMOVE GLASSES, EMPTY POCKETS OF ALL SHARP OBJECTS SUCH AS PENS, PENCILS, ETC.

6. IMMEDIATELY UPON SEEING THE BRILLIANT FLASH OF NUCLEAR EXPLOSION, BEND OVER AND PLACE YOUR HEAD FIRMLY BETWEEN YOUR LEGS.

7. THEN KISS YOUR ASS GOODBYE.

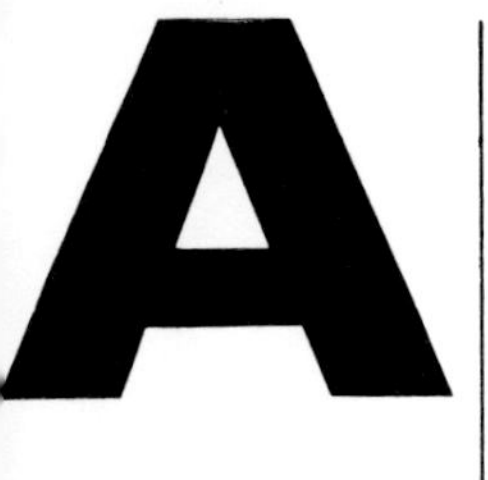

An **Anti-Ballistic Missile** (ABM) system shoots nuclear warheads at oncoming enemy missiles, in 'self-defence'.

An **Atomic bomb** converts matter into energy by the 'fission' of Uranium-235 atoms ($E = mc^2$), releasing heat.

A **Biological Weapon** (BW) is used to spread disease.

A **Chemical Weapon** (CW) is a poison used to kill, immobilize or harass people, or to destroy vegetation.

A **Doomsday weapon** can end all life. Normally a BW, it might be used to blackmail a stronger power.

In an **Electronic battlefield** robots, missiles and pilotless aircraft do all the fighting.

Fragmentation bombs explode pellets that can enter flesh. **Flechettes** are arrow-shaped pellets.

Footfall-activated triggers set off bomb.

Ball bearings are exploded out of container.

Fragmentation bomb buried in ground to this point.

Gamma rays cause radiation sickness. Non-lethal doses can cause **genetic mutations.**

An **H-bomb** (thermonuclear) reaction acts by the 'fusion' of hydrogen atoms and is triggered by the heat from an atomic 'fission' reaction. A 'fission—fusion—fission' bomb also splits Uranium-238 atoms.

ICBMs are **Inter-Continental Ballistic Missiles. IRBMs** have an Intermediate Range.

Russian ICBM with a range of 7000 miles.

Japan was bombed with A-bombs in 1945. At least 80 000 (possibly 200 000) died at Hiroshima, slightly fewer at Nagasaki.

A 20-**Kilotonne** A-bomb was dropped on Hiroshima—equivalent in explosive strength to 20 000 tonnes of TNT.

Leukaemia still claims more than a thousand Hiroshima victims a year—deaths from nuclear radiation.

In a **launch-on-warning** strategy, ICBMs are fired immediately an attack is thought imminent.

A **Megatonne bomb** has the power of a million tonnes of TNT. The USSR has tested a 57-megatonne bomb—the biggest to date.

An **MRV** missile has several (**Multiple**) **Re-entry Vehicles,** each with a nuclear warhead. **MIRV** warheads are Independently-targeted and **MARV** warheads can take Avoidance action against ABM defences.

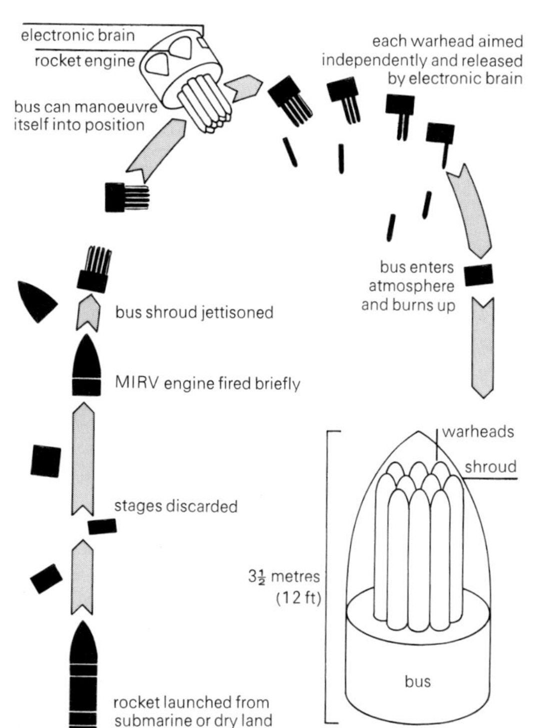

Napalm (jellied petrol) is an incendiary weapon that sticks to skin and flesh as it burns.

Overkill is the capacity to destroy an enemy more than once.

Polaris nuclear submarines have sixteen SLBMs (submarine-launched ballistic missiles). Each SLBM carries three 200-kilotonne MRV warheads and has a 3000-kilometre range.

Poseidon is an improvement on Polaris; each SLBM has ten 50-kilotonne MIRV warheads and a 4500-kilometre range.

The **QU-22B** is a pilotless aircraft being developed for the electronic battlefield.

Riot control weapons, such as rubber bullets, tear gases, squark and strobo boxes, are intended to disperse crowds without causing permanent injuries.

Note the size of this rubber bullet. These are normally fired at the legs.

SAMS are surface-to-air missiles used against enemy aircraft

A SAM missile, one of eight captured intact by the Israelis in Egypt

Strontium-90 is an artificial radioactive substance created in a nuclear explosion. It is carried as fallout.

SRAMs are Short-Range Attack Missiles, launched from aircraft to travel a further 175 kilometres or so.

In a **second strike**, nuclear weapons are used in retaliation, to deny the 'victor' his advantage.

A **tactical 'nuke'** is a small (less than 100 kilotonnes) nuclear weapon for localized fighting.

Trident is an improvement on Poseidon; each has 24 SLBMs with 17 MIRV warheads and a range of 11 000 kilometres.

Uranium-235 is a natural radioactive substance. As it decays, it emits nuclear radiation.

ULMS is an Underwater Long-range Missile System that the US hopes to have operational, using Trident, in the mid 1980s.

U-2 photographic reconnaissance aircraft were used by the US before the advent of 'spy satellites'.

Victory. Used to happen at end of conventional wars. Indistinguishable from defeat in a nuclear war.

The **Warsaw Pact** is the military alliance of the USSR and her allies, opposed to the North Atlantic Treaty Organization (NATO) to which Britain, the US and their allies belong.

X-rays from nuclear explosions can be used as an anti-anti-missile device to confuse ABM radars.

Youth identifiable by a **higher-than-average** level of strontium-90 in the bone, due to fallout from the atmospheric nuclear tests conducted before 1961.

Zones of high radioactivity occur for several days at the centre of a nuclear explosion, and downwind.

Why are there wars?

What is your idea of a caveman? It's probably a bit like the cartoon—an ape-jawed man, with several days' growth of beard, wearing a rough fur tunic and holding a club. Our jokes about cavemen often assume that the club was a weapon used to bash other cavemen or capture women. But all the evidence suggests that cavemen lived very peaceably together, the weapons were used to obtain food. History shows that a man is not a particularly war-like animal: indeed, war as an organized activity is a fairly recent invention.

Even today agricultural communities tend to keep out of wars except when directly attacked. Most people live in relatively peaceful communities and are too busy working for a living to bother with fighting their neighbours. If they did fight, they would have to go back to more or less the same way of living and working afterwards, whether they won or lost. Of course individuals and families have always had disputes. But these are settled *within* a community and don't need wars, armies and war preparations. *Permanent* conflicts of interest between communities and countries only arise when there is trade in raw materials and other goods and hence 'spheres of influence'.

So *organized war*, with permanent armies, developed with trade and the need to protect commercial interests. Communities began to compete with each other and all the members of each community felt a need to band together to protect common interests. A special section of the community was set up to do the protecting and very often became the leaders of the community. Later, these military leaders became so powerful that sometimes they got involved in wars even when they were not protecting the community. But still, the main purpose of every war has been to 'defend a way of life' or 'protect vital interests'. Here is how US President Eisenhower in 1953 explained America's interest in Vietnam:

Now let us suppose that we lost Indo-China . . . . The tin and tungsten that we so greatly value from that area would cease coming, so when the USA votes 400 million dollars to help that war we are not voting a give-away programme: we are voting for the cheapest way that we can prevent the occurrence of something that would be of the most terrible significance to the USA, and our security, our power and ability to get certain things we need from the riches of Indo-Chinese territory.

Life would be simple if these commercial interests were the only reason for wars. If America had given Vietnam a tenth, in aid, of what she eventually spent

on the war, she might have had all the tin and tungsten she needed as well as the friendship of the Vietnamese. In practice, conflicts are confused by racial, religious and political prejudice as well as (some) genuine misunderstandings. Yet, in some countries—notably Brazil—racial prejudice is rare. Catholics and Protestants manage to live together amicably in England and Eire. Communists and Conservatives have co-existed in coalitions in Iceland, Finland and elsewhere. It is clear that conflicts of belief do not lead inevitably to war or violence.

Unfortunately some people need war, or at least the threat of war, to make a living. They fear that disarmament will lose them their jobs. Arms manufacturers and military men often put pressure on governments and make disarmament negotiations more difficult— as Eisenhower warned in 1961:

We have been compelled to create a permanent armaments industry of vast proportions . . . . We must not fail to comprehend its grave implications . . . . In the councils of governments we must guard against the acquisition of unwarranted influence, whether sought or unsought, by the military-industrial complex. The potential for disastrous use of misplaced power exists and will persist.

If Eisenhower was right, society may have to change to ensure that wars no longer occur. We will need to cooperate to resolve conflicts of interest between communities and nations. We will need to prevent anyone having an interest in making, selling or using arms. We can no longer risk war as a means of protecting 'our vital interests'. As Eisenhower has said, 'War in our time has become an anachronism. Whatever the case in the past, wars in the future can serve no useful purpose.'

> God heard the embattled nations sing and shout,
> 'Gott strafe England'—'God save the King'
> 'God this'—'God that'—and 'God the other thing'
> 'My God,' said God 'I've got my work cut out.'
>
> J. C. Squire

Touching off war— the scene at Pearl Harbour on 7 December 1941, when Japanese aircraft made a surprise attack on the main US naval base in Hawaii. Next day America declared war on Japan.

37

The arms trade

The arms trade is very important for the economies of the rich countries. It enables them to sell off weapons once they become obsolete. If this weren't done, the rich countries would not be able to afford to spend so much on new weapons. (Although the US, USSR, France and Britain account for only three-quarters of world military spending, they support almost 95% of the world's military research and development.)

On the other hand, the arms trade can be a crippling burden for the poorer countries. They buy arms to keep up with their neighbours who are also buying arms. For the arms manufacturers this is very good for business. With other commodities, like TV sets and cars, there is a physical limit to the numbers that can be used and the market can get 'saturated'. With arms, the more you sell, the more are needed. The market is never saturated.

Questions (Answers upside down at foot of page)
1. What have South Africa, Nigeria, Israel, Jordan, India, Spain and Greece in common?
2. What have Turkey, Greece, Pakistan, India, Uganda, Libya, Israel and South Africa in common?

Fahmi here to spend £450 M on arms spree

By PATRICK KEATLEY, Diplomatic Correspondent

The Egyptian Foreign Minister Mr Fahmi has arrived in London to ask the Government to agree to a £450 millions deal for the supply of 200 Hawker Siddeley Hawk fighters, 250 Westland Lynx military helicopters, and a range of army equipment.

Answers:
1. Arms from Britain 2. Arms from France.

'This tear-gas spray-gun is just what you need, Madam, to protect you against arms-dealers who accost you in the street.'

'What's holding up deliveries? Both the police and the Mafia are running out of ammunition.'

A stall at the Farnborough Air Show—the Ideal Arms Exhibition?

How it started

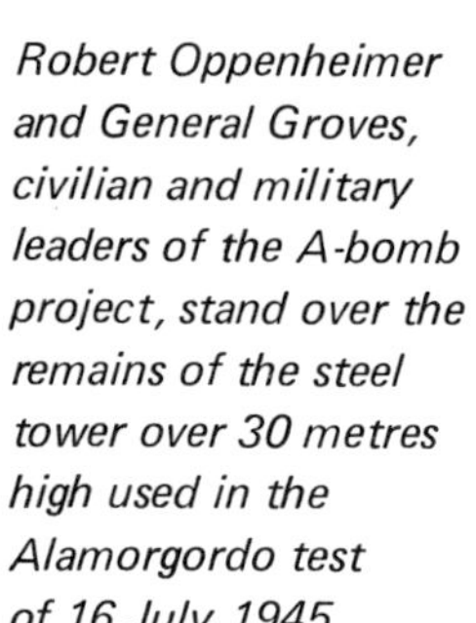

Robert Oppenheimer and General Groves, civilian and military leaders of the A-bomb project, stand over the remains of the steel tower over 30 metres high used in the Alamorgordo test of 16 July 1945.

The nuclear arms race began when the atomic bomb fell on Hiroshima.

At the start of the 1939—45 war, a group of scientists—including victims of anti-semitism like the pacifist Albert Einstein—persuaded US President Roosevelt that the atomic bomb would help defeat Hitler. However, Germany was defeated before the bomb was ready . . .

8th May	Germany surrenders. Japan fights on against US. Soviet pledge to declare war on Japan in three months. The atomic scientists begin to have misgivings.
11th June	Atomic scientists petition for a non-military public demonstration of the new weapon.
12th July	After a series of military defeats, Japan makes peace overtures.
16th July	Atomic scientists poll against military use of bomb. America tests first A-bomb at Alamorgordo.
17th July	Churchill told 'Babies satisfactorily born.' Stalin not told of happy event.
26th July	Truman ignores Japanese peace overtures. US fears that Russia will occupy Japan.
6th Aug.	Atomic bomb dropped on Hiroshima.
8th Aug.	Atomic bomb dropped on Nagasaki. Russia declares war on Japan as promised.
15th Aug.	Big Soviet victories. US still far from Japan. Japan surrenders and is occupied by US.

In 1947, Henry Stimson—the man ultimately responsible for the Hiroshima decision—wrote:

This deliberate, premeditated destruction was our least abhorrent choice. The destruction of Hiroshima and Nagasaki put an end to the Japanese war. It stopped the fire raids and the strangling blockade; it ended the ghastly spectre of a clash of great land armies.

Yet, the decision was taken without support from the Joint Chiefs of Staff. Eisenhower, the Allied military supremo said later, 'It wasn't necessary to hit them with that awful thing.' Today many people believe that it was dropped for political reasons, to ensure that Japan surrendered to America alone: 'the first act of the Third World War with Russia'. Whatever the intention, the use of the atomic bomb had a bad effect on East-West relations. In a triumphal post-war visit to Moscow just afterwards, General Eisenhower sensed Soviet suspicions:

Before the atom bomb was used I would have said, yes, I was sure we could keep the peace with Russia. Now I don't know .... People are frightened and disturbed all over. Every one feels insecure again.

Since 1945 there have been disputes over the future of Europe (especially Germany and Berlin), Asia and the United Nations. In 1949 the Western Powers formed a new military alliance (NATO) and, when a re-armed West Germany joined NATO five years later, the Warsaw Pact was established as a gesture of retaliation. This division of Europe and the world has been profound and there has been an unprecedented arms race. Yet, in retrospect, it is clear that many of the fears that led to the rearmament were not justified. General MacArthur, who was sacked by Truman for being too bellicose in Korea, in retirement said:

Our Government has kept us in a perpetual state of fear—kept us in a continuous stampede of patriotic fervour—with the cry of a grave national emergency. Always there has been some terrible evil at home or some monstrous foreign power that was going to gobble us up if we did not blindly rally behind it by furnishing the exorbitant funds demanded. Yet, in retrospect, these disasters seem never to have happened, seem never to have been quite real.

The Berlin Blockade, 1948—9, created much anti-Soviet feeling. But were the Russians solely to blame? US Secretary of State Dulles confided to journalists, 'There could be a settlement of the Berlin situation at any time. . . . The present situation is however to the US's advantage for propaganda purposes. We are getting credit for keeping the people of Berlin from starving: the Russians are getting the blame for their privations.'

Who's winning

The first major brake on the arms race was agreed
in May 1972 when the US and the USSR agreed to
limit offensive missiles (ICBMs and SLBMs) and ABM
defences. Unfortunately, they did not agree to halt
'improvements' to missiles (such as adding MIRVed
warheads) and have not limited nuclear bombers or
tactical nuclear weapons. In addition, the agreement
does not restrict Britain, France, China, or any other
country.

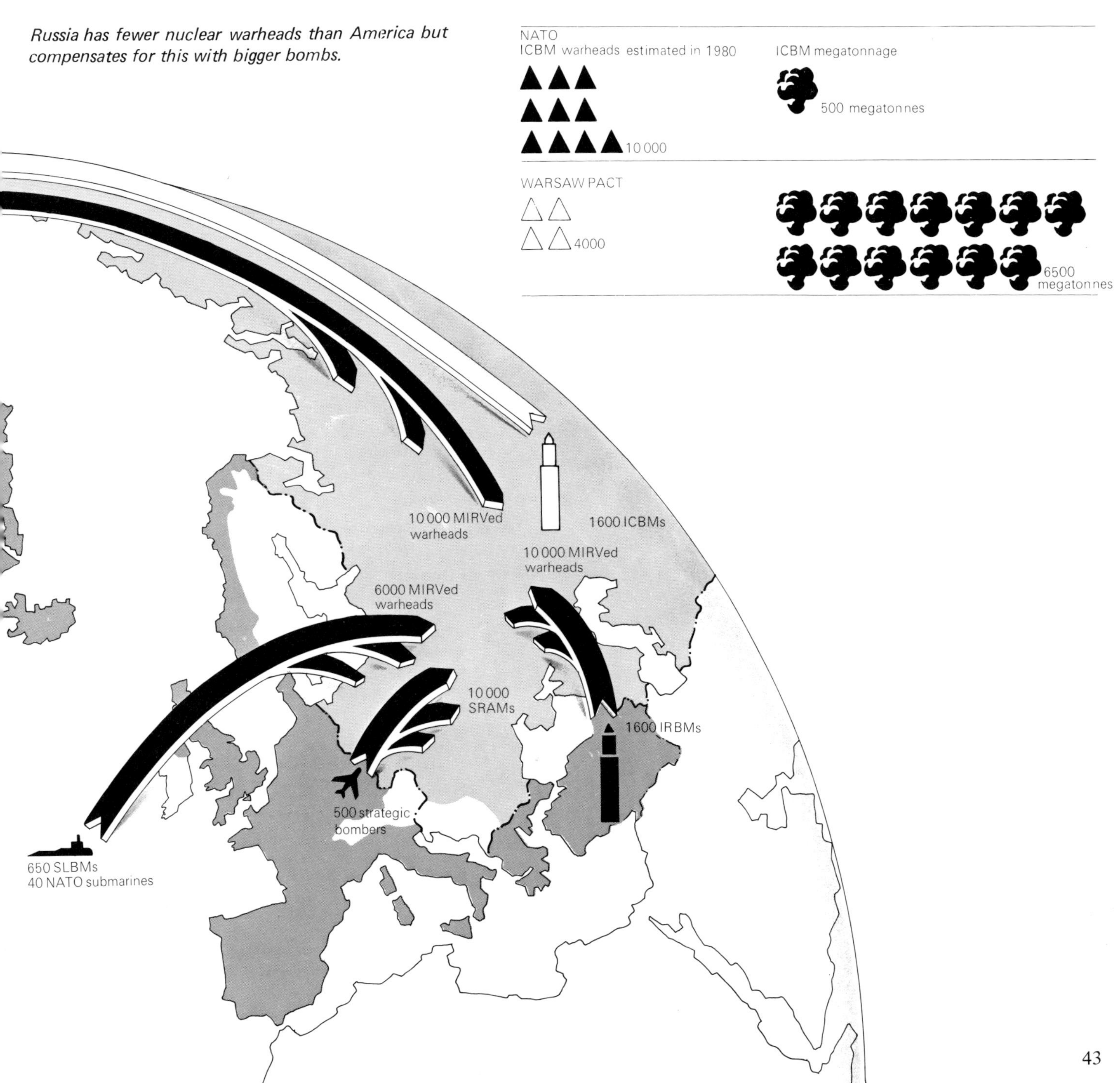

Russia has fewer nuclear warheads than America but compensates for this with bigger bombs.
NATO
ICBM warheads estimated in 1980
10 000
WARSAW PACT
4000
ICBM megatonnage
500 megatonnes
6500 megatonnes
10 000 MIRVed warheads
1600 ICBMs
10 000 MIRVed warheads
6000 MIRVed warheads
10 000 SRAMs
1600 IRBMs
500 strategic bombers
650 SLBMs
40 NATO submarines

Nobody's winning

Which is the most powerful military power? It's not easy to say because the forces are not directly comparable. For example:
* America has fewer ICBMs than the Soviet Union BUT her ICBMs have more warheads

* America has more nuclear bombers BUT fewer armoured divisions.

Comparing like with like can be equally misleading:

* the Warsaw Pact has more armoured divisions BUT NATO divisions are bigger and better equipped

* the Soviet Union has more nuclear submarines BUT America has overseas submarine bases.

The difficulty of comparison leads to many arguments. Are extra Soviet troops worth more than extra American nuclear warheads? Does the Holy Loch base give the US a greater advantage than the USSR gets from its troops in Eastern Europe? Some of those questions are unanswerable. But even if they are answered, time and new circumstances can change the answer. For example, the significance of Holy Loch with 3000-kilometre-range Polaris submarines will diminish when 11 000-kilometre-range Trident submarines become operational.

It is still more complicated when a country has more than one potential opponent. If the US and the USSR demobilized, would this leave Russia in danger from China? (At present most Soviet troops are stationed on the Chinese front.) The only fair disarmament measure would be complete and universal disarmament, since this would leave everyone equal with nothing. But it is feared that one side or the other would obtain a temporary advantage *during* the disarmament process and might be tempted to launch an attack.

It has been suggested that this problem could be overcome by the 'orange-sharing technique' employed by many mothers: one boy divides, the other chooses. Each country should agree to divide up its own territory into, say, twenty areas which it considers of equal military value to itself. The opposing countries then decide which of these areas should be totally disarmed over a fixed period. Thus both sides will have reduced their strength by the same proportion. In practice, the arms race continues although both the big powers are already able to destroy each other completely.

Perhaps after all the detailed comparisons do not matter. In their capacity to destroy one another, the US and the USSR are equal.

This is Revenge, one of Britain's four Polaris submarines. It stays underwater for three months at a time. Should Britain be destroyed, Revenge would take revenge. The three 200-kilotonne MRV warheads of a Polaris missile can do more damage than all the bombing of the 1939–45 war. Each Polaris submarine has sixteen such SLBMs.

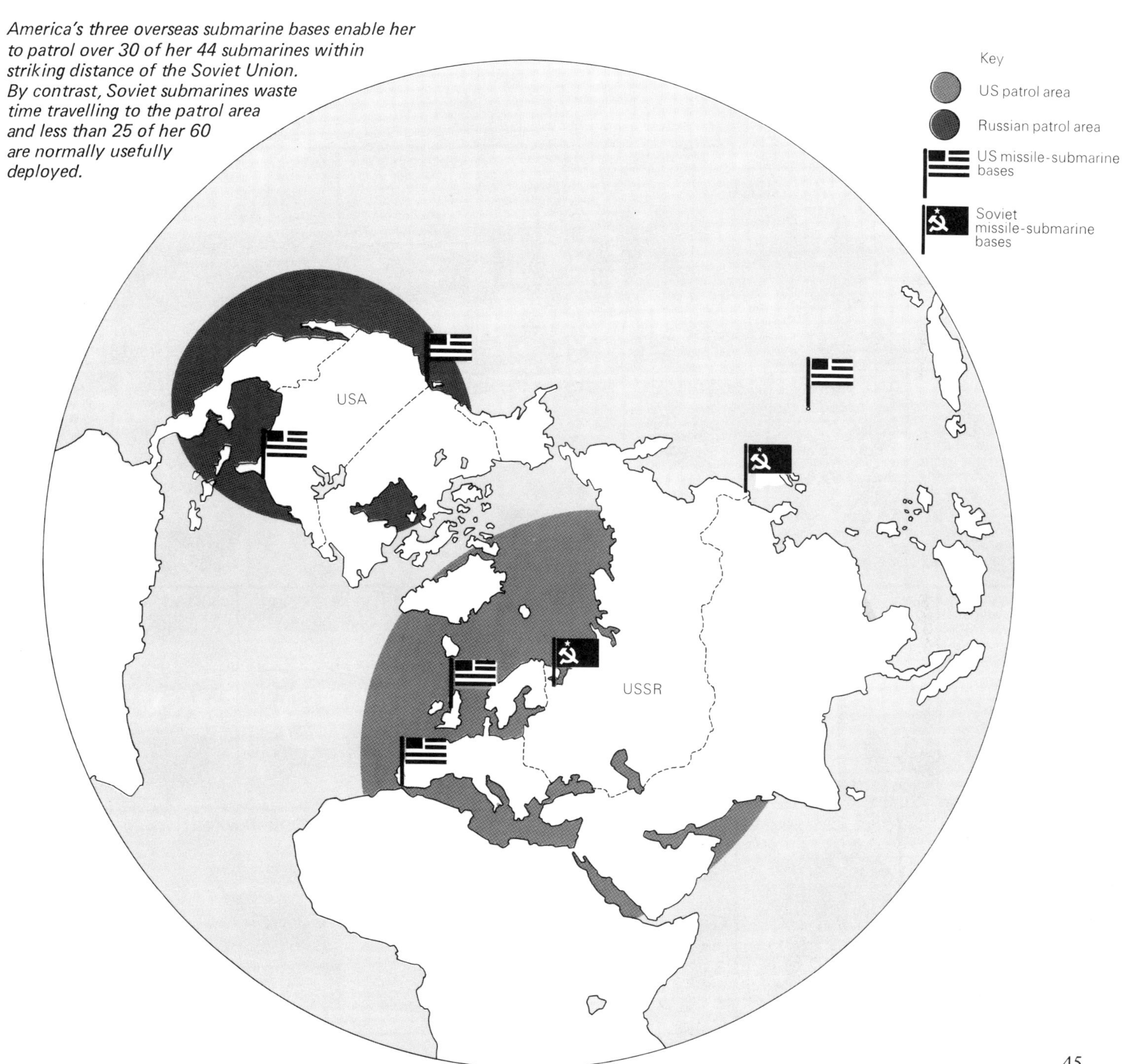
America's three overseas submarine bases enable her to patrol over 30 of her 44 submarines within striking distance of the Soviet Union. By contrast, Soviet submarines waste time travelling to the patrol area and less than 25 of her 60 are normally usefully deployed.
Key
US patrol area
Russian patrol area
US missile-submarine bases
Soviet missile-submarine bases
USA
USSR

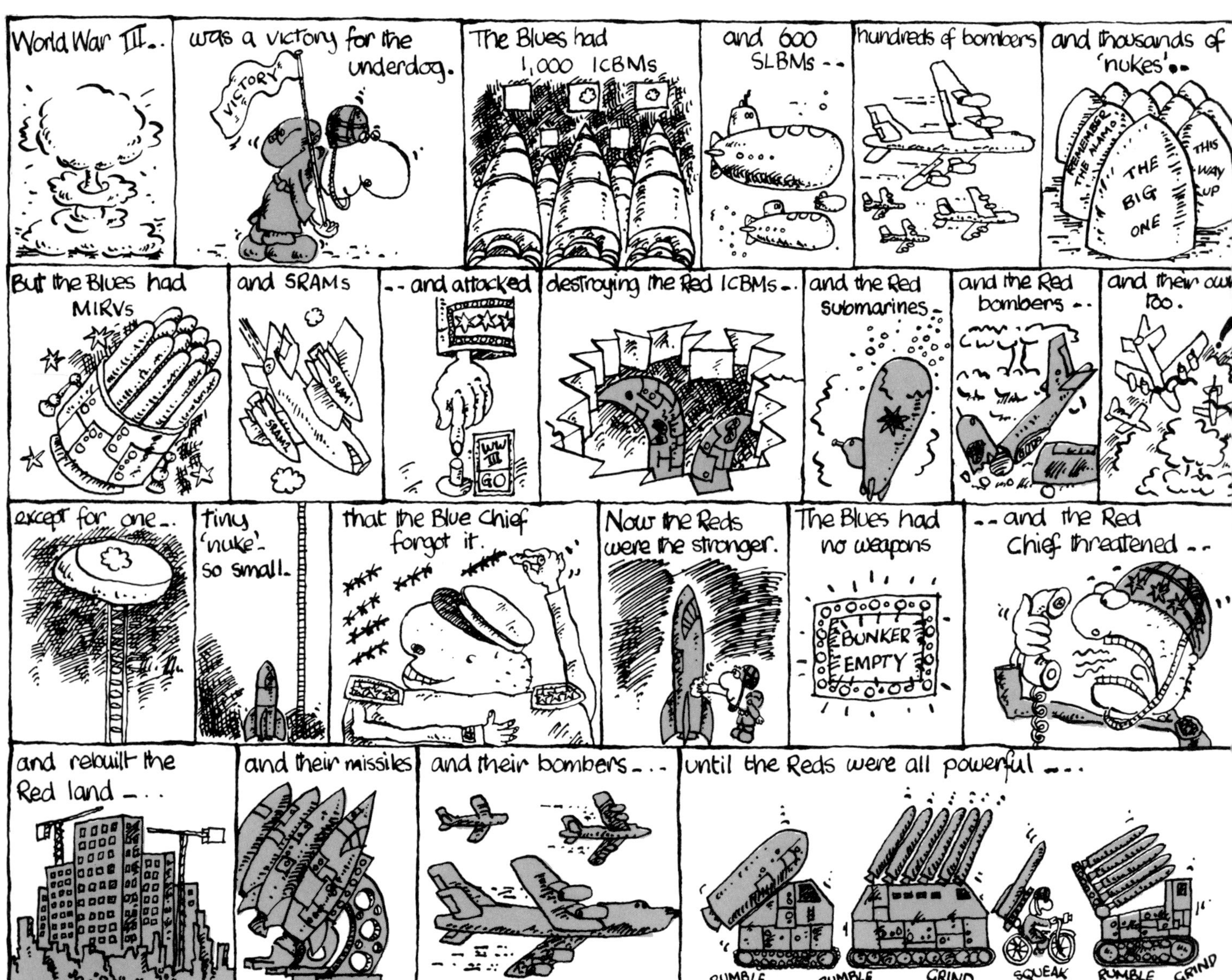

World War III..
was a victory for the underdog.
VICTORY
The Blues had 1,000 ICBMs
and 600 SLBMs --
hundreds of bombers
and thousands of 'nukes' --
REMEMBER THE ALAMO
THE BIG ONE
THIS WAY UP
But the Blues had MIRVs
and SRAMs
SRAM
SRAM
-- and attacked
WW III GO
destroying the Red ICBMs --
and the Red submarines --
and the Red bombers --
and their own too.
except for one -- tiny 'nuke' so small..
that the Blue Chief forgot it.
Now the Reds were the stronger.
The Blues had no weapons
BUNKER EMPTY
-- and the Red Chief threatened --
and rebuilt the Red land --
and their missiles
and their bombers...
until the Reds were all powerful ...
RUMBLE
RUMBLE
GRIND
SQUEAK
RUMBLE
GRIND

Overkill

A few years ago Art Buchwald went round to the Pentagon to talk with spokesman Hiram Hempleweather about overkill. His report makes it very clear why overkill is essential for our security.

Mr Hempleweather was perspiring when I walked into his office.

'It may be too late even to catch up with the Soviets,' he said.

'How's that?' I asked.

'The Soviet ratio of four missiles to our one* could easily change by 1975 to five missiles to our one. We won't have a chance. We have been lulled into thinking that so long as we can kill the Russians once, we have a chance.'

'We don't?'

'Of course not,' replied Hempleweather. 'Suppose we go to the disarmament talks in Vienna and say to the Soviets: "We have enough weapons in our stockpile to kill every Soviet citizen three times" and they come back and say: "So what? We can kill you 15 times." That will put us at a disadvantage.'

'The trouble with your argument,' I told Hempleweather 'is that the Americans don't want to spend money to kill the Russians more than once.'

'Right you are and the Soviets are well aware of this. That's why they're winning the missile race. Some day we're going to discover that the Russians can kill us as many times as they like and there won't be a damn thing we can do about it. All I'm trying to do is to wake up the American people to the fact that in the nuclear arms race it isn't how you are killed, but how many times you are killed, that counts.'

* In his excitement, Mr. Hempleweather was not quite accurate. Under the SALT agreements the numbers of US and Soviet missiles are almost the same whilst the US has more multiwarheads. Precise figures are secret but it appears that the two countries can overkill each other at least ten times and the 'advantage'—if this means anything—rests with America.

Stalemate

So far no-one has dared start a nuclear war. To be successful, the attacker would have to launch a 'first strike' so powerful that his opponent would not be able to inflict serious retaliation. Until now, both Russia and America have had sufficient 'second-strike' capability to inflict 'unacceptable' damage in retaliation—generally thought to be about 20% killed though some experts argue that a 60% death toll could be tolerated.

This nuclear stalemate would be broken if one side or the other developed either an effective 'defensive' system—such as the ABM—or an overwhelming 'offensive' superiority. Neither possibility seems very likely and, with respect to 'defensive' systems, ABMs have been limited to two each by the SALT agreements. Unfortunately, there are no limits on 'improvements' to offensive weapons and in this area the nuclear arms race is proceeding at a furious pace.

Extra warheads are being added to missiles by MIRVing ICBMs and SLBMs. SRAMs are replacing bombs on aircraft. These improvements mean that there is a much bigger danger that someone may risk a 'first strike'. Old-style Polaris missiles are not accurate enough for an attack on military targets, but are all right for retaliation against unprotected civilians. The new-style Poseidon and Trident SLBMs could be used for a surprise 'first strike' with the submarines sneaking close to enemy shores without being detected.

In the meantime, the relative invulnerability of nuclear submarines provides an insurance against a surprise attack by the other side. Let us hope they remain invulnerable! Unfortunately, an enormous amount of money is being spent on Anti-Submarine Warfare (ASW). One idea is for 'hunter-killer' submarines to trail Polaris-type submarines. Another is to sow the ocean beds with acoustic and seismic sensors (as in the electronic battlefield) so that submarines' positions are continuously monitored. A most fascinating project is to train dolphins to track and destroy submarines—the main reason why the US navy does so much research into dolphin behaviour.

Launch-on-warning

Because of the technical difficulties, and the strength of the other side, no-one is likely to start a nuclear war deliberately just yet. But the risk of war by accident grows all the time. The opportunities for human beings to override equipment failure are fewer than ever before.

For example, at present there is no reliable protection for ICBM launching pads, even in underground concrete silos. The simplest and cheapest protection is to fire them all at the start of a war. To reduce delay, the early-warning system will fire the ICBMs directly, without human intervention. But what if the equipment is faulty?

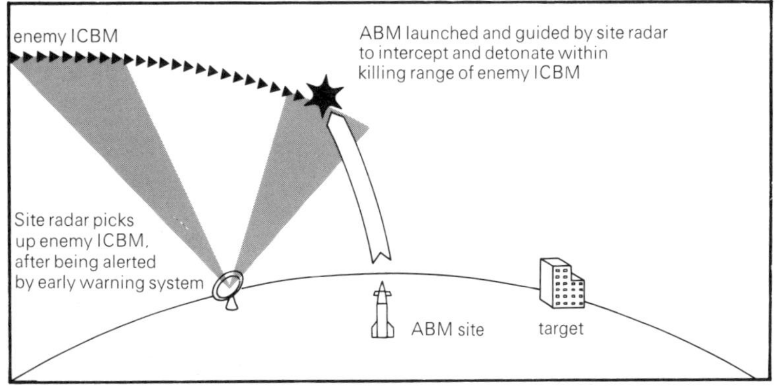

An Anti-Ballistic Missile (ABM) system intercepts oncoming enemy missiles with nuclear warheads. Even a 'near miss' should knock an ICBM off course—though the 'defensive' explosion could wipe out millions of civilians.

An FB—111 carrying SRAMs at a target area, at White Sands missile range, New Mexico.

Super-missile MARV may take over in America's arsenal

From SIMON WINCHESTER : Washington, January 20

After five years of living with the MIRV, the world is about to get something to be known as the MARV. The Pentagon has said it now plans to develop a totally new missile warhead. It will not just contain Multiple Independently Targeted Re-entry Vehicles, but one that is manoeuvrable as well.

Dolphins have been trained to seek out special boats and tap them with their noses. It is hoped they can be trained to destroy enemy submarines this way. One dolphin learnt the trick but refused to perform with explosives strapped to its back.

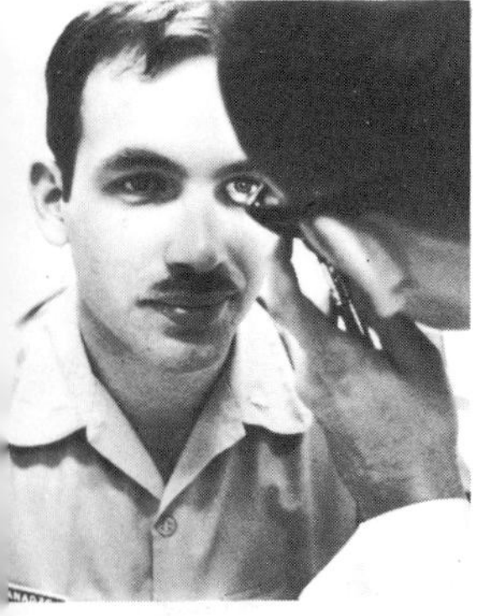

The Home Guard: elaborate electronic scanning devices provide early warning of an air attack.

Medical examinations of missile crews are regular and thorough.

Equipment

On an historic occasion in 1961, US bombers set off in earnest when the NATO early-warning system apparently showed that America was under attack. The bombers were recalled two hours later, still five hours from Moscow. If that same mistake occurred today, Moscow would be destroyed by ICBMs within an hour.

Faults like these are not uncommon but they are usually recognized at once. When ABM and launch-on-warning systems are operational, there will need to be 100% reliability. But defective equipment is not the only danger: signals can be misinterpreted. A moon echo once triggered the early-warning system in Greenland. A flock of geese has been mistaken for a bomber squadron. Such misinterpretations could precipitate a nuclear war. On top of this, there is the risk of accidental nuclear explosions.

Between 1958 and 1968 there were at least 33 accidents involving nuclear weapons in America. In one incident a B-52 jettisoned a 24-megatonne bomb over North Carolina, and only two of the six inter-locking safety devices remained untriggered on impact. If it had exploded it would have been serious for North Carolina and, if mistaken for a Soviet attack, just as serious for everyone else.

Nuclear-laden bombers patrol night and day to guard against a surprise attack. On one occasion a Spanish 'pop' station hit on part of the firing code of an overhead US bomber. (This code was a safety device to prevent the plane crew dropping bombs without authority!) These, and many similar incidents, show that we are at the mercy of sophisticated electronic equipment. If a vital device failed or gave a strange signal in a time of international tension, it might well spark off unintended hostilities.

People

In the quiet town of Hobe Sound, Florida, one night in March 1949, a fire siren sounded and a disheveled man in pyjamas ran out screaming, 'The Red Army has landed!' He was taken to a special hospital where he was described as suffering from 'occupational fatigue'. Later James V. Forrestal evaded his attendants and jumped to his death from a sixteenth-floor window. The US government had to find a new Secretary of State for Defence.

The death of Mr Forrestal, and the life of Adolf Hitler, should warn us that statesmen and generals do not always act coolly and logically. Forrestal's colleagues appear not to have noticed the signs of 'occupational fatigue'—perhaps they were infected with the same paranoia? Can we be sure that only sane people will have the power to start or avoid war?

Politicians are not alone in going round the bend. One in twenty of the crew of a Polaris submarine had to undergo treatment for psychological problems— including a chief petty officer with an acute paranoid schizophrenic breakdown. In a time of crisis it needs only a few irrational acts to precipitate war. In Europe there are thousands of 'nukes' under field command—of Hiroshima-strength or greater—which have to be kept firmly under control. Suppose a field commander went berserk?

The 'Hot Line'

To avoid misunderstandings, the Soviet Union and America have fixed up a direct telephone link so that they can have a quick chat at any time. So if the American President is told of a Soviet attack, he can ring up the Kremlin to see if it's for real. Supposing the answer is, 'No we're not attacking, there must be a fault somewhere', then he can breathe a sigh of relief and tell his aides to ignore the signals. Alternatively, the answer might be 'One of our pilots has gone berserk, it's all right if you shoot him down.' Here also, the 'hot line' can be used to avoid misunderstandings.

It's comforting to know that Russia and America are so alive to the danger of an accidental outbreak of war. Quite clearly both are keen that any future war starts by deliberate intent rather than by mistake.

Nuclear submarine Dreadnought's main control room with space only for two on duty and a visitor.

'Hot-line' conversation from the film *Dr. Strangelove.*

Strike first to save lives

Thermonuclear bombs do so much damage that many people think they should never be used. Hans Bethe, a leading atomic scientist, who helped to make the Hiroshima bomb, has said:

If we fight a war and win it with H-bombs, what history will remember is not the ideals we were fighting for but the methods we used to accomplish them. These methods will be compared to the warfare of Genghis Khan who ruthlessly killed every last inhabitant of Persia.

Those who don't share this view have to decide when their use could be justified. For example, is there any difference between using them first or in retaliation? You may think it obvious that starting nuclear hostilities is worse but others disagree:

We (the goodies) will lose if we don't use nuclear weapons. If we wait for them (the baddies) to strike first, either they do—and lots more on our side get hurt—or they don't—because they can win anyway. So we need to strike quickly so that fewer are killed and good triumphs over bad.

On the other hand, suppose we waited till they bombed us. We would then be too late to attack their bases because they would have fired their missiles already. The most we could do would be to get our own back by killing people in their cities. But it would be too late to win the war.

So the one who strikes first saves lives to win whereas the one who retaliates has no hope of winning and kills out of spite.

In fact no-one expects Britain to become strong enough to launch a successful first strike on her own. Her military dilemma was summed up neatly by Lt-General Cowley of the War Office as, 'Unless we bring the nuclear deterrent into play we are bound to be beaten, and if we do bring it into play we are bound to commit suicide.'

In the late 1950s, Britain supported the NATO policy of 'massive nuclear bombardment of the sources of power in Russia. . . . if Russia were to launch a

A missile crew. Each crewman is equipped with a revolver in case one of the others goes berserk.

major attack. . . . even with conventional forces only'.
The Warsaw Pact rejected this strategy but promised
massive 'counter-city' retaliation if NATO used
nuclear weapons.

Since 1968 NATO has adopted 'flexible response'
strategy. Instead of bombing the 'sources of power in
Russia', NATO would at first drop little bombs
(Hiroshima-strength) on non-vital targets like Prague
or Warsaw. This, it is hoped, would bring the Russians
to their senses and stop the fighting. The trouble
is that Russia might retaliate against non-vital targets
like London or Birmingham—to bring us to our senses.

Maybe Russia is bluffing and is deterred by NATO's
firmness. Maybe we are bluffing but are far too good
and humanitarian to use nuclear weapons if war came.
Maybe. . . .

'Anyone who dreams or talks of winning a nuclear
war is worse than absurd, a menace to his country and
all humanity.' Liddell Hart

'But mate—no one's gonna be mad enough
to start an H-bomb war . . .'

There's no button! Missiles are fired by two keys which must be turned at
the same instant after several other procedures have been completed.

Britain – the third nuclear power

America and Russia have far more nuclear weapons than the rest of the world combined, and were the first two countries to become nuclear powers. Britain achieved this status in 1952 and, since then, France and China have joined the 'nuclear club'. Yet Britain remains in a special position. Although in theory independent, Britain's Polaris secrets were acquired from America. Moreover, Britain, unlike France, cooperates with America in NATO. Consequently, many people who agree with nuclear weapons for America and NATO, see no point in them for Britain.

No-one seriously expects Russia or America to give up nuclear weapons except by mutual agreement. But Britain could do this without significantly affecting the 'strategic balance' and would then join the majority of non-nuclear states in the Western Alliance. Since Britain is too weak to fight on her own in a nuclear war with Russia, and does not expect to fight against France, China or America, it is not clear why an *independent* nuclear force is needed.

If Britain gave up nuclear weapons this would simplify US–USSR negotiations for nuclear disarmament and might help dissuade countries like Israel and South Africa from 'going nuclear'. As an incidental advantage, Britain would be a less important target in a nuclear war. By similar reasoning, Britain might be safer without the NATO bases, of which she has more per acre than any other country. Moreover, as we found in October 1973 when America called a world-wide nuclear alert, these bases mean that Britain can be dragged into a nuclear war without being involved in the dispute or even consulted.

There was very little public discussion at first over whether Britain should make nuclear weapons. As it was a security matter, military men discussed the issue privately before telling the politicians. Eventually, after they had decided that Britain did need

'Proper little trouble maker, aren't you?'

atomic bombs (1948), they told the Prime Minister. He agreed with them but the rest of the government was not told until later. It was 1951 before the news was made public.

Since then many people have opposed this decision. One million British people signed the 1951 Stockholm Appeal calling for a ban on nuclear weapons. After 1958, when the Campaign for Nuclear Disarmament (CND) was formed, there have been many marches and demonstrations. The CND Easter marches from the Aldermaston Atomic Weapons Research Establishment, and the CND symbol, have become known all over the world.

Despite these protests, government policies have not changed—though the Labour Party has promised on several occasions to close all nuclear bases. Campaigning against nuclear and other weapons still continues.

The CND symbol was designed for the first Aldermaston march (in 1958) by Gerald Holton. It blends ancient symbols for human suffering and despair with the modern semaphores for N and D—the initials of Nuclear Disarmament. The central motif is a human being in despair; the circle represents the world and the unborn child, both threatened by nuclear war. The symbol is meant to be in white against a black background representing eternity. Since 21 February, 1958, it has appeared throughout the world as a symbol of peace, sometimes in other colours but always with the semaphores essentially unaltered.

55

Arms control and disarmament

Signing on . . . a press agency photo shows US President Ford and Soviet Communist Party Leader Mr Brezhnev concluding another round of disarmament talks with another signing ceremony.

Signing off . . . US Secretary of State W. P. Rogers signs the ceasefire agreement to end the war with Vietnam.

'The use of nuclear and thermonuclear weapons is a violation of the Charter of this organization.'
UN General Assembly, 1961

Partly in response to world-wide campaigning for disarmament, international relations have thawed since 1961. One proof of this is the seven important multilateral treaties agreed. Taken as a whole, these may slow and could even halt the arms race. Certainly, without these treaties the world would be in an even worse state.

The **Partial Test-Ban Treaty** (PTBT) bans nuclear weapons' tests everywhere but underground. The US, USSR and Britain signed it in 1963, but the other two nuclear powers (France and China), who had hardly got started at the time, believe the PTBT to be a dodge to keep the first three ahead. France and China have not signed and so continue to pollute the atmospnere with nuclear fallout.

The **Non-Proliferation Treaty** (NPT) prohibits the spread of nuclear weapons. The non-nuclear powers have agreed to this *provided* that the nuclear powers (under Article VI) work seriously for nuclear disarmament. Although signed by 92 countries (10 May 1975), several near-nuclear countries (notably South Africa, Portugal, Spain, Israel, Pakistan, India and Argentina) have refused, claiming that the NPT is a dodge to keep the nuclear powers ahead.

The convention prohibiting biological weapons provides for the destruction of BW stockpiles. So, whilst all the other treaties merely limit arms growth, the **BW Convention** is a genuine measure of disarmament. It is interesting that it has no arrangements for verification: the good faith of the signatories is considered sufficient guarantee that it will be effective.

The four other treaties restrict areas of military activity. The **Outer-space Treaty** prevents weapons of mass destruction being sent round the moon, in orbit round the earth or anywhere in space—and has been signed by 69 countries, including the Vatican. Simi-larly, the **Sea-Bed Treaty** stops countries leaving nuclear bombs lying around at sea—and has been signed by 46 countries, including Switzerland. The remaining two provide for 'nuclear-free' zones on land. You will be fairly safe from nuclear attack if you live in **Antarctica** or **Latin America**. Unfortunately, the Latin-American country most able to make its own nuclear weapons—Argentina—has not signed this treaty or the NPT.

Bilateral US—USSR agreements

As required by Article VI of the NPT, the nuclear powers have been negotiating amongst themselves. The US and the USSR agreed in 1972 to limit their ICBMs and SLBMs to the numbers shown on pages 42 and 43. They have installed 'Hot Lines' to reduce the danger of accidental war or misunderstandings. But it is doubtful whether the non-nuclear countries will be satisfied with these agreements. There are no limi-tations on 'improvements' (MIRVing ballistic missiles or SRAMing aircraft loads), and such developments are potentially more serious than the initial construc-tion of ICBMs and SLBMs. By the early 1980s, both the US and the USSR could be close to possessing a 'first strike' capability. Their agreement in 1979 to 'limit' weapons actually allows each side to develop new weapon systems.

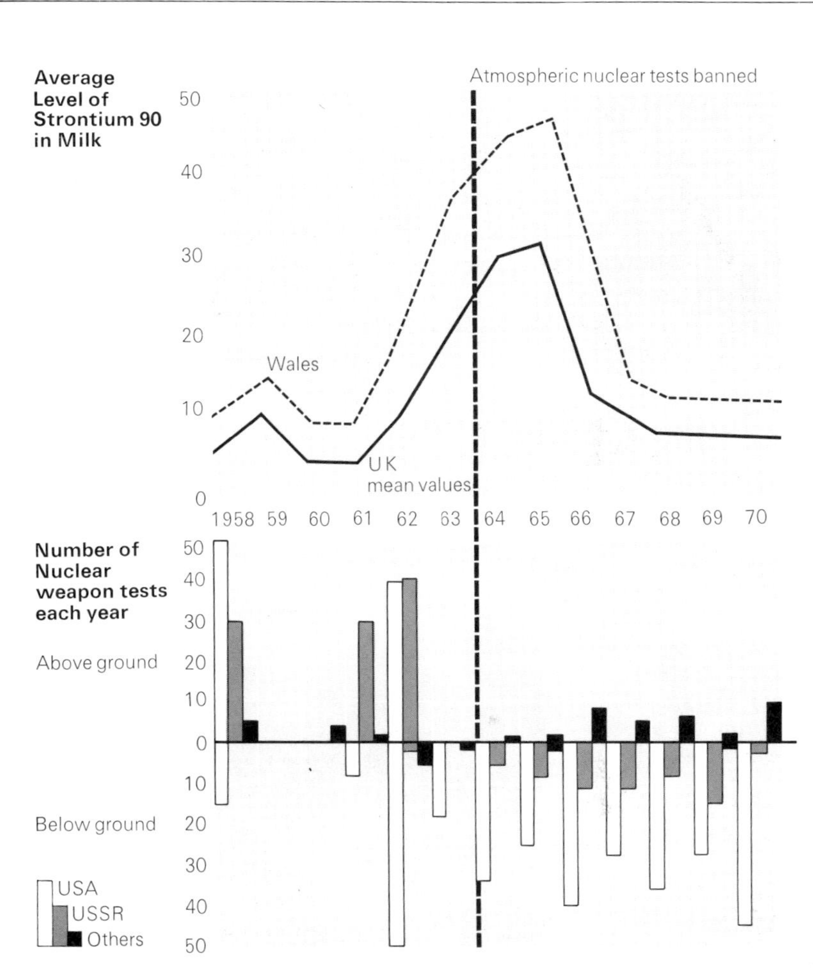

Areas with high rainfall and calcium deficient soil accumulate the most strontium 90. For example, in March 1963 Merionethshire milk contained 74 strontium units—double the Welsh average which itself was well above the U.K. average. In 1956, when the hazard was first recognized, the usual value was 4.

Growing children retain about a quarter of the strontium they ingest as food. Those born between 1958 and 1968 may have more in their bone than the 10 strontium units 'safe level' originally recommended by the Medical Research Council.

The moratorium on nuclear testing (1960—1) slowed strontium 90 pollution but, anticipating the 1963 Partial Test Ban Treaty, more than a 100 atmospheric tests took place in the next 2 years and this caused the biggest-ever increase.

James Bond is alive and well and working in Surbiton

Not all artificial satellites are launched for space exploration or scientific research. At least a third are 'spy satellites' used for photographic reconnaissance. The James Bond of today is far more likely to work in a Surbiton office, interpreting aerial photographs, than taking his own and living it up with Olga.

Since 1963 photographs from a height of 24 kilometres have been able to pick out 100 mm painted strips and 300 mm square objects. It has thus become simple to observe building sites and deduce how many missiles, submarines, aircraft and other weapons are under construction, their sizes, where they are aimed at and other important military information. Nor is this photography confined to daylight hours. Infra-red pictures show variations in temperature, revealing movements of lorries and trucks at night. Electronic reconnaissance can pick up radar signals and other sources of electromagnetic radiation.

Combining that data yields far more than can be deduced from daylight photography on its own: for example, a missile silo emplacement with grass growing on top can be pinpointed from its thermal emission, reflectivity and odour characteristics. The first Soviet SAM delivered to North Vietnam was correctly recognized as a dummy without supporting gear—despite its camouflage!

Camera-carrying satellites observe weather developments—or military activity.

This night-time infrared photograph of a Texas airport even shows the exhaust heat from the aircraft on the runway.

Radio communications are monitored all the time. America's National Security Agency, at its unit in Grosvenor Square, London, is known to listen to low-level communications in Whitehall—including the radio traffic of the Foreign Office and MI 6. So there are few secrets between enemies—or friends. Everyone has a good idea what everyone else is doing—a good thing as it reduces the risk of misunderstanding.

Spying can also help disarmament as it enables both sides to have confidence that agreements are being honoured. For arms control, or for disarmament, it is necessary to inspect only as often as is needed to discover significant changes. These changes may take months, involving major movements of men and equipment. Satellite reconnaissance is more than adequate to check on such movements. So, when the US and the USSR made the SALT I agreements, verification was left to 'national technical means' —in other words, spy satellites.

Underground nuclear testing

This cannot be checked simply by satellite reconnaissance. In this instance, seismic detection is necessary to supplement aerial inspection of suspect activities. Underground nuclear tests were not banned at the same time as atmospheric tests because America claimed that seismic detection was not good enough. They said that earthquakes and small nuclear explosions could not be distinguished.

Yet, since 1963 when the PTBT was signed, underground nuclear explosions in the 5—10 kilotonne range have been easily detected by seismic stations around the world. As all meaningful tests are from five to five thousand times stronger than this, it is obvious that the problem of verification is not the real reason why there is no comprehensive test-ban treaty.

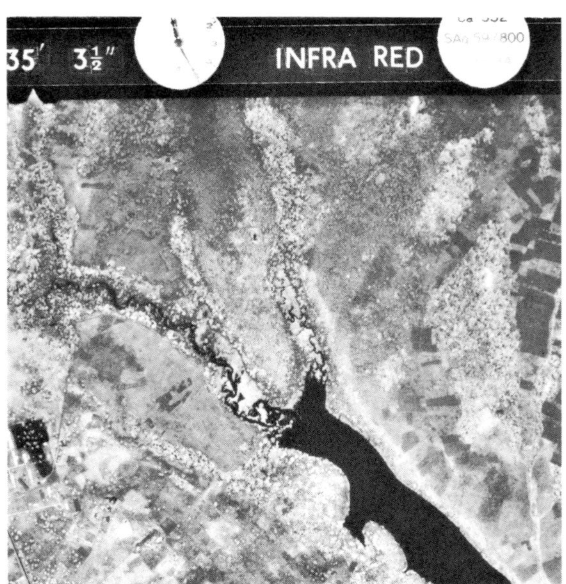

The details of infra-red photographs show a skilled interpreter if any unusual activities are going on.

The discovery of an American U2 plane on a spying mission over Russia stopped the summit conference on disarmament.

The law of the jungle

It was the ELEPHANT that called them together. He was tired of finding his mid-day meal ruined by the trampling of yet another battle—and somewhat concerned at the disappearance of so many of his peace-loving plant-eating companions. He proposed that fighting be abolished.

The LION said, 'You can't change animal nature; fighting is natural.' He did agree however that some animals were cruel and unnecessarily destructive. In the days when he had been all-powerful, fighting was heroic and chivalrous and the more skilled with tooth and claw always won. He deplored the modern tendency to launch attacks from the air. 'Beaks and wings should be abolished.'

The YELLOW DRAGON sneered, pointing out that the LION was protected by his four nit-picking bird companions who had stood guard over him ever since the sad day when the LION lost his teeth. The COCKEREL agreed and, flushed with his sucess in a recent battle with a worm, proposed that everyone should be free to do his own thing. This pleased the SNAKES and SKUNKS no end as most animals told them off for using CBW, but upset the DOVES.

The YELLOW DRAGON supported the COCKEREL, wishing him success in his efforts to forge an alliance with the LION against the BEAR and the EAGLE. The YELLOW DRAGON denounced all the others as running dogs and paper tigers. There was an uncomfortable silence. Until recently few had realized that the DRAGON really existed and his fire-breathing made them hot under the collar.

The EAGLE doubted whether it was possible to halt progress. Improved methods of fighting were inevitable. His prey were always finding new ways of hiding under trees and in holes—so why should he forego the advantage of surprise attacks from a great height? As for the accusation of cruelty, this was bound to happen whilst trying out new techniques. The only solution would be to abolish tooth and claw—under adequate control and supervision of course—so that air attacks would not be needed.

Amid the clamour, the BEAR proposed general and complete disarmament. He was the main rival to the EAGLE and had been growing in strength and influence ever since routing the EAGLE's double-headed cousin. The plant-eating creatures, although suspicious of his power, often followed his lead. They cheered his proposal to abolish teeth, claws, beaks, wings, tusks and fangs. 'After that,' continued the BEAR, 'we could all embrace one another.' His rivals remained unimpressed.

MAN, who at that time was an insignificant grubber, took no part in the discussion. What they need, he mused, is someone to keep the peace for them. That same night he invented the club.

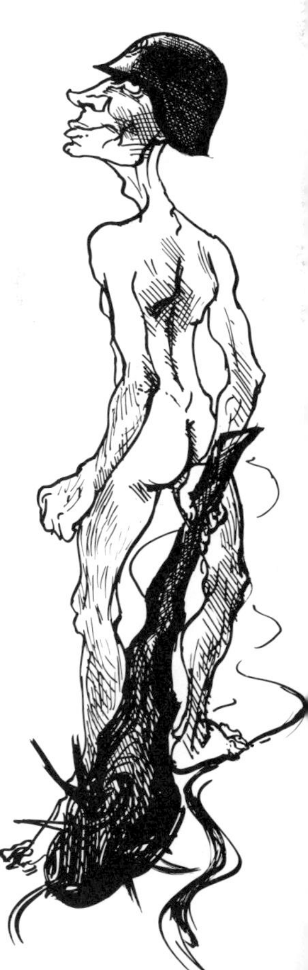

Wars will cease when men refuse to fight

During this century thousands of pacifists have gone to gaol—and in some countries have been executed--for refusing to fight. If enough people did this, war would be impossible. But, for every pacifist, there have been hundreds of others only too willing to take up arms.

Most religions condemn warfare so it's surprising there aren't more pacifists. Early Christians, and other religious groups, would not kill under any circumstances. In the Second century, St Cyprion preached,

The world is wet with mutual bloodshed and homicide is a crime when individuals commit it but is called a virtue when many commit it. Not the reason of innocence but the magnitude of savagery assures impunity for crimes.

The test for pacifists comes when someone like Hitler is born. Millions of Jews and political dissidents were being murdered until Nazism was defeated by war. An absolute pacifist (**conscientious objector**) contends, even in this situation, that the evil of killing is absolute, corrupts the executioner and hence cannot be condoned even in defence or to stop genocide.

A **political objector** is prepared to fight but only if he agrees with his country's war aims. Before the First World War, the European socialist parties all agreed that they would not fight for their respective governments because, they believed, there would be nothing but suffering in it for ordinary people. (In the event, most forgot the promise when war broke out —but that's another story.)

Although many Germans refused to fight for Hitler, and thousands of Americans would not fight in Vietnam, most people—and most religious leaders— support their government, 'right or wrong'. This blind patriotism prevents rational thought and intimidates pacifists. Very few people oppose a war once it is under way. For example, in 1956 the Labour Party was strongly opposed to British intervention at Suez. Yet, when troops landed, the Labour Party leader wished them every success!

Nuclear weapons have made pacifism much more acceptable. Pope John said, 'It is impossible to conceive of a just war in a nuclear age.' Eisenhower agreed, 'Whatever the case in the past, war in the future can serve no useful purpose. People holding this view may be termed **nuclear pacifists**.

'I have never been a complete pacifist and have at no time maintained that all who wage war are to be condemned . . . some wars are justified, others not. What makes the peculiarity of the present situation is that if a great war should break out, the belligerents on either side and the neutrals would be all, equally, defeated. This is a new situation and means that war cannot still be used, but only by a lunatic. Unfortunately, some people *are* lunatics, and, not long ago, there were such lunatics in charge of a powerful state. We cannot be sure this will not happen again . . .' Bertrand Russell

American pacifist Tom Dunphy has travelled the world to spread his message that 'War is a racket'. He is seen in his costume as General Wastemoreland.

'We must learn to accept war as a child or we will never do it'.
Kathleen Riley

'You will not end wars by stopping children playing at soldiers, although you would probably stop children playing at soldiers if you prevented war.'
Lord Hailsham

Is science to blame?

Are the scientists to blame for making the weapons described in this book? Or should we blame politicians for ordering the work to be done? Or does the fault lie with us all, for leaving things to the politicians and not taking responsibility for the consequences of our work? Although the scientist has a special responsibility, everyone contributes in some way towards this situation.

A scientist operates in a world where hundreds of millions of pounds are spent every year on military research, where scientific endeavour is, in practice, dominated by military requirements. Although science is in theory free to develop in any direction, this enormous investment ensures that a large proportion of the work of university science departments, government laboratories and industrial firms has direct or indirect application to war and weaponry.

Against this background, no scientist can ensure that his work is used only for peaceful purposes.

Arthur Galston, an American scientist who made his reputation through research into hormones, found that his discoveries enabled others, twenty years later, to devise new weapons for use in Vietnam. A devoted campaigner against CBW, he has described himself as 'the last of the scientific innocents'. A scientist may choose to work on medicines, fertilizers or pesticides but he cannot be sure that his work will not have other, unexpected, applications.

Take the nerve agents. Some of the more recent ones were the result of work in this country by an industrial concern looking for better insecticides. They told us about this and they were looking of course for really low toxicity, but found rather to their disgust that some of the agents they were using had rather high toxicity. We, in fact, investigated, told the Americans about them as part of our liaison, and the Americans, you now know are making these in ton quantities for offensive purposes.'
Eric Haddon of CDEE, Porton, speaking in the film, *Plague on your Children*.

Given sufficient ill-will and ingenuity, almost any scientific discovery can be developed as a weapon. This led the world-famous British scientist J. D. Bernal to comment,

Because the weapons which are being prepared have largely been forged by science, the scientist has a special responsibility and should be deeply concerned in all efforts to stop war and to remove its polifical and economic causes.

But this responsibility does not end with the scientists. Scientific research, humanely applied, has the power to change the world for the better, just as much as research into weaponry has changed it for the worse. This is everyone's concern:

Above all I would like to believe that the people in the long run are going to do more to promote peace than are governments. Indeed I think that people want peace so much that one of these days governments had better get out of their way and let them have it.
Eisenhower